D0857277

EVERYONE'S TRASH PROBLEM: NUCLEAR WASTES

EVERYONE'S TRASH PROBLEM

NUCLEAR WASTES

MARGARET O. HYDE
AND BRUCE G. HYDE

McGRAW-HILL BOOK COMPANY

NEW YORK · ST. LOUIS · SAN FRANCISCO · AUCKLAND
BOGOTÁ · DÜSSELDORF · JOHANNESBURG · LONDON
MADRID · MEXICO · MONTREAL · NEW DELHI · PANAMA
PARIS · SÃO PAULO · SINGAPORE · SYDNEY
TOKYO · TORONTO

Library of Congress Cataloging in Publication Data

Hyde, Margaret Oldroyd
Everyone's trash problem—nuclear wastes.

Bibliography: p. Includes index.
Summary: Discusses nuclear waste, a by-product of nuclear energy with the potential to cause serious and irreversible damage to the environment.
1. Radioactive waste disposal—Juvenile literature. 2. Atomic power-plants—Waste disposal—Juvenile literature. [1. Radioactive waste disposal. 2. Atomic power plants. 3. Pollution] I. Hyde, Bruce G., joint author. II. Title.
TD898.H92 614.8'39 78-23859
ISBN 0-07-031551-5

23456789 MPMP 83210

For
Eric Rudolph Strebel
Ralph Joseph Strebel
and
Stephen Curt Strebel

Other books by Margaret O. Hyde

Addictions: Gambling, Smoking, Cocaine Use and Others
Alcohol: Drink or Drug?
Brainwashing and Other Forms of Mind Control
Fears and Phobias
Hotline
Know About Drugs
Know About Alcohol
Mind Drugs
Speak Out on Rape!
VD: The Silent Epidemic
What Have You Been Eating?
Where Speed Is King (with Edwin)

Books by Margaret O. Hyde and Edwin S. Marks

Mysteries of the Mind
Psychology in Action

CONTENTS

1 NUCLEAR WASTES: HANDLE WITH CARE

When you see bumper stickers that say SPLIT WOOD, NOT ATOMS; STOP NUKES; or GO NUCLEAR OR DON'T GO, do you feel confused? Why do people become so emotional about nuclear wastes?

Just imagine a football field that is piled 6 feet high with trash.

Then consider a four-lane highway stretching from the East Coast of the United States to the West Coast covered 1 foot deep with nuclear trash. This is about 1 billion tons of trash.

These are varying estimates for the year 2000 of the amount of nuclear wastes. No one expects the nuclear wastes to be lying around in either place. As trash volume, the projected amounts are not especially large. In fact, one of the advantages of nuclear energy is the small amount of waste produced. But the dangers of radioactivity make this small amount something that must be handled with care—so much care that the final solution to permanent storage of nuclear wastes is many years away.

Nuclear waste cannot be disposed of like ordinary trash. To throw it away, burn it, compact it, dump it, or drop it at sea are common ways to dispose of unwanted materials. Recycling works well with paper, aluminum cans, car bodies, and some other materials. In a sense, just about all trash is recycled eventually. The trash that is buried in the soil travels by little rivulets to the rivers or is absorbed by plants and eaten by animals. In one way or another, most materials are used again and again. Why not recycle the nuclear wastes? The special problems involved in recycling are discussed in Chapter 7.

What should be done with the millions of gallons of nuclear wastes that are stored in tanks in the ground on a temporary basis? What should be done with the spent or used fuel that is being stored temporarily in pools of water near commercial power plants?

Many volumes have been written about the possible answers to these questions, and suggestions are plentiful. If you ask a dozen different people what should be done with present and future nuclear wastes, you may be answered in a dozen different ways:

"Shoot them into space."

"Dilute them by scattering them far and wide."

"Send them to Antarctica."

"Recycle them."

"Bury them in the ocean."

"Put them in glass and bury them."

"Bury them in salt mines."

"Put them in ceramics and bury them."

"Put them all in one place and cover them over with a heavy lead shield."

"Nuclear wastes are a time bomb. No way will ever be found to dispose of them safely."

"The technology exists now for safe storage."

"No program will ever be acceptable to everyone."

Some of the suggestions that seem unusual have been seriously considered by scientists who are working on the problem.

So far, concern about nuclear wastes has been confined mainly to scientists and environmentalists. A few farmers have discovered problems in their fields because nuclear wastes have leaked there—but not many farmers. Some wastes have leaked onto the ocean floor, but not many people have been affected by it. Headlines over newspaper articles announcing the discovery of leaks in the Pacific Ocean in 1976 were small, but 61,800 55-gallon drums, containing items such as wiping rags and overalls worn by those handling nuclear materials, were involved. May future leaks be larger or more serious?

Nuclear wastes are accumulating in the United States and around the world. Some will be hazardous for longer than any human culture has lasted. Controversy exists about whether we have the technology to build nuclear-waste disposal facilities that will be safe for at least 10,000 human generations. Herman Krugman, a scientist at Princeton University's school of engineering, suggests: "It may turn out that the predictive capabilities of man do not suffice for such a judgment [about how to safely store wastes] to be made on a solid basis."

Certainly the subject of nuclear wastes is important to this and future generations. Even if all nuclear power production and national defense nuclear programs stopped at this very minute, the problem of existing wastes would not be solved.

If nuclear fuel provides about 30% of the total United States electricity demand by 1990, many new reactors will be adding to the supply of nuclear wastes. The average reactor can provide the same amount of electricity as that from 10 million barrels of oil or 3 million tons of coal, and it will produce about 30 tons of spent fuel each year. Each source of electricity has its own problems, and one of the major concerns about nuclear energy is spent fuel rods and the depleted uranium they contain. These rods are being reprocessed in some countries, but in the United States they remain in temporary storage pools while we consider how to proceed.

Should the United States reprocess fuels and recycle materials in spent fuel rods? Should you be concerned if a nuclear storage site is planned for your state?

To understand the many problems involved, you must have some understanding of atoms, radioactivity, how nuclear power is produced, breeder reactors, the difference between low-level and high-level waste, and other scientific information. The basic ideas are not complicated. Isolating nuclear wastes is both complicated and urgent.

2 THE ATOMS THAT MAKE NUCLEAR TRASH

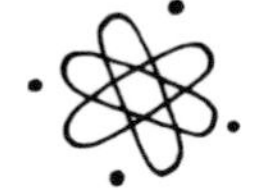

Nuclear wastes, like everything else in the world, are made of atoms. They must be handled with care because of the vast number of atoms in nuclear wastes that are breaking apart. These atoms are radioactive, so they can cause harm to the human body.

DANGER! RADIOACTIVITY may be a familiar sign to you. Signs warn of danger in hospitals, research laboratories, agricultural stations, art museums, and any number of places where radioactive atoms are put to work. Tremendous amounts of shielding protect workers and the public in nuclear power plants, where there are large amounts of radioactivity. The number of atoms breaking apart at such places is beyond anyone's imagination. The action of these single atoms must be considered in protecting ourselves against radioactivity.

You cannot see an atom because it is so small. Scientists estimate that there are more than 100 million atoms spread over the head of a pin. One cannot really appreciate such minuteness. If each atom in a grapefruit measured 1 inch in diameter, the grapefruit would be as large as the earth. One atom is about a millionth the width of a human hair. It would take a row of 100 million copper atoms placed side by side to stretch an inch.

In spite of the small size of atoms, scientists have been successful in seeing individual atoms through the use of advanced electron microscopes that enable them to analyze atomic actions. Highly trained atomic scientists can focus on a single atom and separate it in a picture from thousands of surrounding atoms.

Atoms are amazing because of their small size, but they are amazing for other reasons, too. The more one learns about them the more intriguing they become. Consider their basic structure. Not all atoms weigh the same. This is true of the atoms in this page even though the paper looks the same all over.

Based on the way atoms act, scientists have been able to compare the weights of atoms and place different kinds in a chart, beginning with the lightest (hydrogen) and ending with the heaviest. Uranium, which is famous as a fuel for nuclear reactors, has the heaviest atoms that occur naturally. Some man-made atoms are heavier and have properties that create some of the problems in nuclear waste.

If you could look at a single atom, what might you see? Pretend that you can magnify the atom so that it is the size of a large room. The room-sized atom is mostly empty space, but there is a speck in the center, roughly the size of a fly. This is the nucleus, the heavy part of the atom. The nucleus is about 100,000 times as small as the whole atom, and it is from this minute part that atomic or nuclear energy is released. The outer part of an atom is composed of one or more particles known as electrons, each of which has a negative electrical charge. (An electric current is a flow of electrons.) Electrons orbit around the nuclei of atoms. The idea of electrons orbiting around a nucleus somewhat like a planet orbiting around the sun is not entirely precise scientifically, but it helps to use the image in order to give a rough idea of what is happening.

Hydrogen, which is the lightest atom, is also the simplest. It has one proton, a positive particle, composing the nucleus, and one electron in orbit. All other atoms have neutrons in the nuclei. These neutral particles, which weigh just a trifle more than protons, are important in any discussion of nuclear energy. When neutrons are released from the nuclei of atoms they play a part in continuing the action inside a reactor, which provides the heat for the production of electricity in nuclear power plants.

The tiny nucleus of the atom is a very important and complicated structure. No one knows exactly what is going on in the nucleus even though whole libraries of data have been compiled about it.

There is one example that shows how complex the nucleus is. Just as the earth spins on its axis, neutrons and protons spin on an axis. The earth turns around once every 24 hours. These unbelievably tiny particles spin around about 30 billion trillion times per second, in a circular path, which may be horizontal, vertical, or tilted. They travel at various speeds, and the orbit may be circular or eliptical.

Two main forces are at work in the nucleus of an atom. One is the strong, or nuclear, force by which neutrons and protons all attract one another. The other is the electrical force by which protons repel one another.

The nuclei of atoms must be split before atomic energy can be released. Think of the nucleus of an atom as a package in which the contents are tightly bound. Some kinds of atomic nuclei seem to be more tightly "wrapped" than others. Perhaps some combinations of atomic pieces fit together better than others. Certainly some kinds of atomic nuclei can be made to break apart more readily than others. Some break apart naturally; these atoms are the radioactive atoms. Radioactivity is the property of the nuclei of the atoms of certain elements that break apart bit by bit, releasing waves and particles according to a certain pattern.

Radium and uranium are examples of atoms that split without prodding. This is not true of all kinds of uranium, however. Most of the uranium that exists in nature is in two forms. Although these forms of uranium behave the same chemically, they have different numbers of particles in their nuclei and their atoms do not weigh the same. Uranium-235 has 92 protons and 143 neutrons in the nucleus (add them and you get 235) and is less stable than uranium-238, which has 92 protons and 146 neutrons and which exists in nature in much larger amounts. Of every 1000 parts of uranium, only seven parts are uranium-235.

No one knows exactly why the nuclei of certain atoms are unstable so that they split naturally. According to one theory, protons and neutrons in nuclei are arranged in concentric shells.

"Magic number" nuclei have completely filled outer shells. The ratio of protons to neutrons in a nucleus appears to play a part in whether an atom is stable or radioactive.

Scientists do know that certain kinds of atoms are unstable. They split apart spontaneously, and when they split apart they release radiation. Natural radioactive elements go through several generations of splitting. Some of the atoms that are formed from the original fission are radioactive, and they are called "daughters" of the original substance. These daughters may produce daughters, and so on, until they end eventually as stable elements.

This splitting of atoms makes them valuable for the production of energy. Radioactivity has been put to work to improve the quality of life in many fields, such as agriculture, medicine, and industry. Whole books are filled with descriptions of processes and techniques that make use of splitting atoms. But as use increases, so does nuclear trash. And the splitting atoms in nuclear wastes make this a special and dangerous kind of trash.

3
WHAT IS YOUR ANNUAL RADIATION DOSAGE?

Radioactivity from splitting atoms is not something far away from you, but is all around you as part of your natural environment. You cannot see, feel, smell, hear, or taste radioactivity except in very large doses. Radioactivity is not new. The whole earth is radioactive to a small degree; it always has been; and it always will be. You cannot completely escape radioactivity. It is the amount and kind of radioactivity around you that is important. Everyone who wants nuclear wastes stored far from home is aware of this.

Cosmic rays from the sky, dental and medical X rays, and radioactivity from splitting atoms damage cells in the human body to some degree, but not everyone agrees on how much exposure it takes to cause serious damage. People have always had some exposure to background radiation from the splitting atoms that are found naturally on earth. How much of an increase is safe? Whether anything above zero exposure has an important effect is the subject of controversy. Damage to health depends on the total dose the body receives, whether the dose is local or general, and what kind of tissue is involved.

How much radioactivity are you exposed to in a year even if nuclear wastes remain isolated from you? You can compute your annual dosage to some degree, using measurement units such as millirems and millirads. For most purposes, these units can be used interchangeably. A millirem is a unit that measures the estimated biological effects of radiation on your body and a millirad is a unit of the amount of energy absorbed by body tissue. "Milli" is a prefix meaning one-thousandth, and even a whole rem or a whole rad is a very small amount of radiation.

To count your radiation dosage, first consider where you live. For cosmic rays at sea level, record 40 millirems, and add 1 for every 100 feet of elevation. Cities along the coast are considered to be at sea level. Some typical elevations are Dallas, 435 feet; St. Louis, 455 feet; Minneapolis, 815 feet; Atlanta, 1050 feet; Spokane, 1890 feet; Las Vegas, 2000 feet; and Denver, 5280 feet. At

Denver, the average exposure to cosmic rays contributes 93 millirems to your count.

The kind of building you live in adds a small amount of radiation; strange as it may seem, people who live in tents receive less radiation from their housing material than those who live in stone houses. Wood houses add 35 millirems, concrete and brick add slightly more, and stone walls increase the count from 50 to 100 millirems. Multiply the appropriate number by the fraction of time you spend at home.

The earth itself supplies some background radiation, but the great majority of people receive only 15 millirems per year from the ground. In some parts of Colorado, India, and Brazil, concentrations of uranium, thorium, and radium in the earth increase the amount of background radiation.

Your annual food, water, and air intake increases your count by 25 millirems. Most of the natural radioactivity in food comes from carbon and potassium in the plants and animals you eat. Pollution causes more radioactive atoms to be present in some water than in other water. Even discounting pollution from industrial sources, radioactive elements such as radon and radium may increase radioactivity in water. At one time such water was thought to have curative properties and people gathered at hotels near springs to drink the "medicines."

In 1978 Thomas F. Gesell and Howard M. Prichard of the University of Texas school of public health announced that household use of water naturally high in radon may be exposing people in certain areas to an increased risk of lung cancer due to radon gas liberated from the water.

Unless you live near springs that supply radium water, you need not be concerned about the minute amount of radioactivity in what you drink. The danger of water becoming polluted by nuclear wastes is extremely small at present.

How you live helps determine the amount of radioactivity to which you are exposed each year. If you travel frequently in jet planes, you increase your count. Add 4 millirems for every 6000 miles you travel. Count the average number of hours you watch color television a day, and add 1 millirem per year. If you watch black-and-white television, divide by 2. (Radiation varies with individual sets and conditions.)

Medical X rays are often a big factor in radiation exposure. Doses for X rays vary a great deal depending on the equipment, how it is used, and the area of the body being X-rayed. New standards that follow the principle of "as low as reasonably achievable" (ALARA) are being encouraged for use in nuclear industries and in medicine and dentistry. No one can set a maximum permissible dose for a patient, since the physician must weigh the benefits against the risks for the patient.

With increased knowledge about the effects of radiation on the human body, protective shields have been put into use in medical and dental procedures which involve radiation. You may have worn an apron that contained lead while having your teeth X-rayed. Technicians who work with X rays, radium, and other radioactive materials watch the patient who is being examined or treated from a room where they are protected from frequent exposure.

The average exposure from medical and dental techniques each year may be from 35 to 90 millirems. Of course, many people are exposed to much larger amounts and some are exposed to none. A chest X ray may deliver from 1 to 20 millirems to the bone marrow, which is sensitive to radiation (too large a dose may lead to leukemia). The dose from a chest X ray may vary from 10 to 3000 millirems. But the X ray may be essential in determining treatment that can save the patient's life. For example, radiation from radium can destroy cancer tumors. So it is with many medical techniques.

If you have had a chest X ray this year, add 100 or 200 millirems. Add 20 for a dental X ray, and 2000 for an X ray of the gastrointestinal tract.

If you live near a nuclear power plant, distance plays a part. At the site boundary multiply the average number of hours per day at home by a factor of 0.2; if you live a mile away decrease the factor to 0.02; at 5 miles, to 0.002. Thus at the site boundary, if you are home 12 hours a day, add 2.4 millirems per year. These estimates come from the American Nuclear Society, which claims you need add nothing if you live more than 5 miles from a nuclear power plant.

Adding millirems from all these sources gives you almost the complete annual exposure to radiation. Another possible source is radiation from fallout.

There is still some radioactivity in the air and soil from the fallout of atomic and hydrogen bombs that exploded years ago. Some of these atoms continue to enter and leave your body, adding about 4 millirems to your natural background radiation.

In 1978 China was the only nation in the world that continued open-air nuclear tests. Open-air tests are believed to cause the highest amount of radioactivity in the environment. The most serious recent crisis in cloud-borne nuclear wastes occurred in September 1976 when an atomic weapons test in China (small by modern standards) sent radioactive dust particles skyward. They were carried eastward by prevailing winds.

Cows in New England were eating grass contaminated by particles that had traveled halfway around the world. The particles were carried across the Gobi desert, over the Pacific Ocean, and, within a week, across the United States, where they fell with the rain on the pastures where cows ate the wet grass containing radioactive iodine. Cows are considered highly efficient in "vacuuming" fresh fallout from the landscape; radioactive elements got into their milk. Radioactive iodine in cow's milk is carefully monitored during fallout alerts, since children, born and unborn, are especially susceptible to its effects. Radioactive iodine tends to concentrate rapidly, especially in the developing thyroid glands of the young, and to cause cancer.

Another element of special concern after a bomb test is radioactive strontium. It concentrates in cow's bones and milk and in the bones of those who drink the milk. Radioactive cesium, presumably from fallout, has been found in some mallard ducks at a level 2000 times greater than the level in their food. While cows can be taken from pastures after a fallout alert and confined to barns until the levels of radioactive material decline, no one knows where all the fallout goes and what is a safe amount in practice.

What is known is how long it takes for half the original amount of a radioactive substance to decay. The iodine in the fallout decays so rapidly that half the original amount is gone after only 8 days. For the strontium-90 and cesium-137 in fallout, half of the original amount decays in about 30 years.

For some radioactive wastes the half-life is very long. A radioactive isotope of carbon contains 14 protons and neutrons in its nucleus, and therefore is called carbon-14. If you could count out 100 atoms, you could expect 50 of the atoms to decay in 5568 years. It would take another period of 5568 years for half the remaining atoms to decay, and so on. Some radioactive atoms have half-lives of a billion years.

You can get an idea of how much radiation you are exposed to each year by comparing it with the average for a person in the United States, which is about 225 millirems. But you cannot be sure that you have calculated your dose accurately or that it will stay the same year after year.

HELPING TO STOP ADDITIONAL FALLOUT

This is a partial list of organizations working to control the spread of radioactivity through the control of nuclear testing as well as other means. You may be interested in writing to them for literature.

Critical Mass
P.O. Box 1538
Washington, DC 20013

SANE
318 Massachusetts Ave. NE
Washington, DC 20002

Scientists' Institute for
Public Information
355 Lexington Ave.
New York, NY 10017

League of Women Voters
1730 M St. NW
Washington, DC 20036

Union of Concerned
Scientists
1208 Massachusetts Ave.
Cambridge, MA 02138

4 HOW MUCH RADIATION IS TOO MUCH?

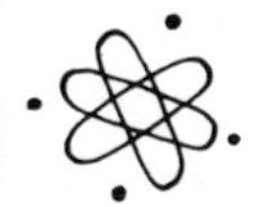

The problem of nuclear wastes would be simpler if the question *How much radiation is too much?* could be answered simply. Obviously you have been exposed to some radiation from birth, even if you have never been in the sunshine. Some of the rays from splitting atoms are present in your body.

Radiation has been defined as energy moving through space in invisible waves. The frequency of these waves determines their effect on you and other living things. Non-ionizing radiation comes from the sun, cosmic rays, microwave ovens, and other sources. Life could not exist without heat and light radiation from the sun; as you probably know, overexposure to sunshine can cause damage to your skin: sunburn or even cancer. Ionizing radiation includes X rays and four kinds of radiation from splitting atoms: alpha, beta, and gamma rays, and neutrons. X rays are similar to gamma rays but come from other sources. Some kinds of rays penetrate more deeply than others, and can cause more damage.* As ionizing radiation passes through body cells, it knocks electrons away from their atoms, which can cause changes in the ways the cells function.

These changes may not be evident for decades. This makes it difficult to gauge how much damage a given amount of radiation will cause unless the radiation is so intense as to cause death or disease in a matter of hours or weeks.

Many people have been exposed to rather large doses of radiation and lived. What humans eat and drink varies, their environments vary, their genetic susceptibilities vary, and so on, so what is true of one person may not be true of another. A dose that is "safe" for one person may cause any number of health problems for another. Just suppose you have a sister who has asthma, for example. Her chances of developing leukemia are greatly increased by an extra 100 millirems of natural background radiation. Vari-

*See Glossary: Alpha particle; Beta particle; Gamma rays

ous diseases compound risks. The national watchdog, the Bureau of Radiological Health, says that adverse effects from ionizing radiation are directly related to the amount of radiation.

Long before the bad effects of radioactivity were determined, women were employed to paint radium on the dials of watches and airplane instruments by hand. Many of them were in the habit of making points on their brushes by twirling them on their tongues, and in doing so they absorbed a good deal of radium into their bodies. Radium tends to settle in the bones, and the first case of bone cancer in these women was discovered within five years. The health of about 5000 of the workers was followed thereafter, and cancer of the bone developed in about 28% of the women. They probably had received huge doses, perhaps more than 1.2 million millirads. In spite of this, 72% of the women did not develop cancer. The great variation in humans plus the fact that it may be necessary for multiple factors to act together or in sequence to cause cancer may have protected them.

Today workers in nuclear plants are monitored by radiation badges that record their exposure and by other instruments that help health physicists to maintain records of exposure. How much is a "permissible dose" for workers is still being argued. There have been a number of reductions in this dose for occupational workers and the public during the past 35 years.

Many nuclear plants, when they need repair work done in high radiation exposure areas, employ temporary workers. Since radiation dosage is cumulative, it is better for workers to be exposed for only a short time. Since workers often move from one area to another, or from one country to another, they need to cooperate so the results of their exposures can be monitored.

Increased awareness of our lack of knowledge about the biological effects of radiation has led to many studies of workers and other people exposed to unusual amounts of radiation. In 1978 records were examined of men who worked during the infancy of the nuclear submarine program at the Portsmouth, New Hampshire, naval shipyard. Cancers have long latent periods, and radiation damage would not show hazards until many years after exposure. According to a study by Dr. Thomas Najarian of the Boston Veterans Administration Hospital, the cancer death rate for employees at the shipyard was 38.4% while the national average is 18%.*

Also in 1978 there was renewed interest in the

**Medical World News,* March 20, 1978, p. 21.

work of Dr. Thomas F. Mancuso. He had been studying cancer rates among workers at the government's Hanford, Washington, nuclear laboratory for about ten years when his research contract was canceled in 1974.

Since then a number of studies have been made of these workers, but findings differ. The differences have been blamed on interpretation. Two studies were presented at the American Association for the Advancement of Science meeting in 1978: both showed a higher than expected incidence of cancers of the pancreas, of blood plasma cells, and of the lungs among the workers. Opponents of these studies remind the public that one must consider other occupational exposures of these workers such as exposure to chemical and other hazards before their employment at Hanford.

Veterans of the "Big Smokey" atomic tests of 1957 are being studied by the Center for Disease Control in Atlanta. The Defense Nuclear Agency has set up a tollfree number (800-336-3068) and urges the 300,000 military personnel and civilians who took part in the tests to contact them. This is the result of reports indicating a possible increase in the number of cases of leukemia among those who were exposed to increased radiation during the atomic testing. Only about 25,000 people reported by phone or by letter in the first several months. This demonstrates some of the difficulty in making reliable studies of what happens to people who are exposed to unusual amounts of radiation.

Early in 1978 the Department of Energy announced plans to expand its studies of the health of laboratory workers who have been exposed to radiation. Medical follow-ups of personnel of the department and its contractors are included. It is intended to include anyone who was exposed to more than 5 rems in any year since 1942.

Monitoring has developed a great deal since the end of the 19th century, when the "mysterious lung disease" was known among uranium miners in Europe before radioactivity was discovered. Later, the disease was diagnosed as lung cancer and its cause was traced to radon gas, a radioactive product of uranium decay.

More recent studies of men who work in uranium mines in the United States have shown a wide variation in their exposure to radioactivity. In well-ventilated mines exposure was only 0.5 rad per year, while in unventilated mines it was 100 rads. Relatively few people work in uranium mines, but some information from these studies is important to many of us. It was discovered that dust and cigarette smoking increased the risk of lung cancer in mine workers by five to ten times.

Naturally occurring radioactive substances such as polonium have been suggested as a significant factor in the increase of lung cancer among cigarette smokers. It may be that small amounts of radioactive substances are trapped by the fine hairs of tobacco leaves. Radioactivity in cigarette smoke may work in combination with dust and chemical air pollutants to play an important role in causing cancer. Whether or not you wish to accept this risk is a personal matter.

Your chances of living in an area where radioactive materials were used as fill before homes and schools were built over them are extremely low, and are even lower now that efforts are being made to do something about such wastes. Back in the days when nuclear-waste disposal was largely ignored by the public, some health problems were created by radioactive tailings (sand).

The first step in the production of nuclear energy is the mining and milling of uranium ore. When the rocklike ore is crushed at the mill and uranium is removed, a slightly radioactive, gray sand remains. There are piles of tailings scattered over many areas near uranium mills. In Grand Junction, Colorado, people carted some of this sand away and used it as construction fill for the foundations of houses and other buildings. Radioactive decay of elements in the fill produced radon gas, which seeped up through cellar cement slabs and collected inside houses. Some tailings were used under streets, driveways, sewer lines, and swimming pools, but the most dangerous situations were in houses where the radon was not free to circulate in open air. As a result, about 5000 homeowners in Grand Junction received letters in 1971 warning them that a study had confirmed the presence of uranium tailings on their properties. The radiation exposure level was so high that corrective action was necessary: the radioactive tailings were removed from beneath the foundations at government expense.

An awareness of the danger of similar materials used at other sites developed. An estimated 140 million tons of radioactive tailings have been left exposed at abandoned mines in the western United States. Tailings from a uranium mine should be kept from entering the environment of people; they must not be used as fill or allowed to seep into streams and rivers where they could eventually reach water supplies.

Even more recently, exposure from mill tailings has come to light. In the summer of 1978 radioactive contamination was found at a number of sites where atomic energy had formerly been explored. At Canonsburg, Pennsylvania, an industrial park has been built in an area where about 1000 tons of waste were buried in 1957. A survey showed that about 120 workers in the area were being exposed to low-level radiation which might increase their risk of developing cancer. At the United States Marine Corps training center in Middlesex, New Jersey, radioactivity was traced to a uranium ore-processing plant that formerly occupied the site. Wastes from the plant were buried on land now occupied by a Catholic school. Later in the summer of 1978, the federal government set in motion the legislation needed to clean up these and other sites where uranium mills have left materials that continue to pose radioactive hazards.

Today many people are asking how much exposure to radioactivity is too much. Dr. Karl Z. Morgan, a famous health physicist who was director of the health physics division of the Oak Ridge National Laboratory from 1943 to 1972, says: "An overwhelming amount of data shows that there is no safe level of exposure and no dose of radiation so low that the risk of malignancy is zero. The question is—how great is the risk?"* But he also observes that the risk is one of probability, just as there is a chance of an accident every time you ride in a taxi, and that there probably is no occupation that does not

*Karl Z. Morgan, "Cancer and Low Level Ionizing Radiation," *Bulletin of the Atomic Scientists* (34, no. 7), September, 1978, p. 30.

pose some risks to its workers.

When dealing with risk, people have a very difficult time deciding how much risk is acceptable. No one lives with zero radiation, and everyone seems to accept the risk of background radiation without any problem. Many people who live in California have no problem accepting the risk of an earthquake even though there is a fair probability that a major earthquake will occur there within 20 years, and many scientists believe it is virtually certain that a major quake will occur there within 100 years. Most citizens of San Francisco appear simply to dismiss the possibility from their minds. Perhaps they feel there is little they can do about the problem. This need not be the case with protection from radioactivity from present and future nuclear wastes. Understanding the problem is a good beginning.

One thing that makes the nuclear-waste problem so difficult is that some of the materials are man-made and our experience with them is relatively brief. Plutonium, which is formed at the end of the process by which uranium fuels commercial power plants, hardly existed until the atomic research of the war in the 1940s. The newness and the long half-life of plutonium make it a very special kind of nuclear waste.

5
PRODUCING NUCLEAR WASTES

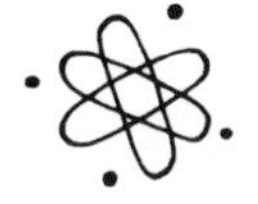

If nuclear weapons programs already produce so much waste, should you be concerned about wastes from commercial power plants? Probably you will join the chorus of "Don't bury the wastes in my backyard" if you live near a nuclear power plant, although you may receive less radiation than if the plant were fueled with coal. Radioactive materials in coal are released into the air when it burns. However, the ashes from a coal-fired plant cause much less of a disposal problem than the wastes produced in a reactor.

Some people say there are comparatively few wastes from commercial power plants compared to the wastes from military programs. In a way this is true, but though the volume of military wastes is large, the concentration of radioactive material in them is small compared to that in commercial wastes. In 1977 the amount of radioactivity in commercial spent-fuel waste was comparable to that in military waste and the commercial waste was increasing more rapidly.* This is one reason that what is done with the wastes from commercial plants is everyone's trash problem.

The processes by which uranium fuels nuclear power plants and by which nuclear wastes are formed are mysterious to most people. Actually, there is nothing mysterious about the basic procedures in nuclear power plants. There are different varieties of reactors, but the principle remains the same. Heat from fissioning, or splitting, atoms is put to work to produce steam. The steam is used to drive turbines that generate electricity. The reactor takes the place of a fuel box in a power plant that uses coal, oil, or gas to make steam.

Enough atoms must be splitting at all times in a nuclear reactor to keep the chain reaction in motion and produce enough energy to produce electricity. Imagine an atom of uranium-235 being hit by a slow neutron. Since it is a radioac-

*Report of the Nuclear Energy Policy Study Group, *Nuclear Power: Issues and Choices*. Cambridge, Mass.: Ballinger, 1977, p. 244.

tive atom, one with an unstable nucleus, there is a high probability that the neutron will be captured by the nucleus. When this happens, the nucleus becomes so unstable that it very quickly splits. Some energy is released, some neutrons are released, and two or three kinds of atoms are formed. Some of the newly formed atoms are radioactive and continue fissioning.

By working with concentrated amounts of uranium-235 scientists learned how to produce a controlled chain reaction. In any amount of uranium there are some free neutrons. If enough uranium-235 is brought together, the free neutrons hit the nuclei of enough uranium atoms to start a chain reaction. The uranium atoms that split free more neutrons; these hit more nuclei of uranium atoms that split, releasing still more neutrons, and so on. The energy released along with the neutrons is mainly in the form of heat. On transferring the heat to water, the water becomes steam that drives turbines to generate electrical energy.

Enough atoms must be splitting to keep the chain reaction in motion, but too much activity produces too much heat. This could melt the core, the part of the reactor where the action is taking place. Elaborate safety precautions are taken to prevent this.

Consider what happens in many nuclear

power plants. Uranium fuel has been enriched to increase the proportion of U-235 to U-238 so that a chain reaction can be maintained. The fuel, in the form of pellets, has been loaded inside rods that are ½ inch in diameter. These pellets are uranium dioxide, a compound of uranium. This is a tough ceramic substance with a melting point of 2746° C (5000° F). Pellets are sealed in long tubes and the tubes bundled into metal frames to make up a fuel assembly. A large number of fuel assemblies held in a grid make the core of the reactor. The core is held inside a large steel tank known as the reactor vessel. When the reactor is ready for action, cooling water will circulate around each fuel rod carrying heat away.

You cannot turn a switch to start the splitting of uranium atoms any more than you can turn a switch to stop them, yet the amount of fission must be controlled. When the fuel rods are in position, control rods are placed so that they absorb unwanted neutrons that are constantly being released from the fuel. Control rods are made of materials such as boron and cadmium, materials that act something like blotters to absorb unwanted neutrons.

When the loading is finished, a set of control rods is withdrawn. Now more neutrons are free to dart about inside the reactor vessel. More control rods are gradually withdrawn. The rate of withdrawal is carefully monitored by neutron-counting instruments. When the rods are at a certain position, the reactor becomes "critical." This is the point at which the chain reaction becomes self-sustaining. The control rods, also known as safety rods and regulating rods, are adjusted to maintain the desired operating level.

As neutrons and fission products collide on a massive scale with surrounding material, their kinetic energy is converted into heat. Heat would be a great problem in the operating of a nuclear reactor if arrangements were not made to remove some of it. If a reactor were operated at a high power level, a level at which there was much activity, there might be enough heat to melt the core.

In a water-cooled reactor, water surrounds the fuel rods and acts as a coolant by absorbing heat from the rods. Heat is carried away from the core of a reactor by the water under pressure to an exchanger where other water is heated, turning to steam that is used to produce electricity.

In addition to control systems and coolants, reactors need moderators. Sometimes the coolant serves as moderator in addition to carrying away the heat; sometimes other materials are used. In either case, the moderator aids fission by slowing down the rate of motion of neutrons. Strange as it may seem, slow-moving neutrons are more effective in triggering fission than fast ones. Collisions slow the speed of neutrons, but some may be "wasted" by being absorbed by fission products if too many collisions are involved before the neutrons reach the speed at which they best trigger fission. Water, heavy water, graphite, and beryllium are among the materials that are used as moderators. They slow down neutrons without absorbing many.

A reactor has been described as a sort of three-dimensional, high-speed game of pinball, cooled by a fan and controlled by rods. The game is played in the fuel area, with the pinball or neutron speed being controlled by other balls, the moderator atoms. And it is all carefully wrapped to protect those who play the game and to keep it from damaging itself.

Nuclear reactors comprise a most serious and dangerous game, of course. Today there are many kinds of reactors using various materials as fuels. But no matter what the fuel or type of reactor, fuel is enclosed in a number of safety systems, which are built according to elaborate specifications with backup systems in case of human error or mechanical emergency. Even so, there are many people who are concerned about the possibility of accidents.

By the late 1980s there may be 500 commercial reactors operating in 30 countries. Some fuel rods must be replaced each year in order to continue efficient functioning; although each reactor produces a relatively small amount of nuclear trash, about one-third of the rods are changed in the core of each reactor each year. It is not that they no longer contain any usable fuel. Only a small part of the fuel is consumed in any cycle of operation.

The fuel atoms that split, or fission, release new kinds of atoms along with the neutrons that cause more fissioning, and these nuclear ashes are contained in the fuel rods. Some of the ashes absorb neutrons that are needed to continue the chain reaction, thus slowing down the reaction. If fission products were not removed after a certain time, they would stop the reaction. A typical nuclear reactor that has been in operation for many months has several hundred pounds of fission products distributed among many tons of fuel in the rods that make up the core of a reactor.

Most fresh fuel rods in use are not highly radioactive: most of their radiation is from the uranium pellets, a kind of radiation that penetrates only a short distance and can be stopped by a piece of paper. But some of the nuclear trash in spent fuel rods is so hazardous that spent fuel rods must be handled by remote control from the time they are removed from the reactor until they are stored. Another difference between fresh and spent fuel rods is their temperature. Fresh and used fuel rods look very much alike, but spent fuel rods are hot from the heat of splitting atoms. They are temporarily stored in concrete basins, which are open at the top and filled to a depth of 40 feet with water. The water circulates around the spent fuel rods, removing heat and acting as a shield by stopping neutrons from traveling from one fuel rod to another. This is a safety precaution against a possible chain reaction in the storage area, where, of course, one is not wanted. Although the fuel rods that have accumulated unwanted fission products are not as likely to start a chain reaction as those with fresh fuel, this could happen if precautions were not taken.

As mentioned earlier, storage pools at many nuclear power plants have been filling up with spent fuel rods. This has been true in the United States since commercial reprocessing stopped in 1972. Some storage pools have been enlarged. When and if there is reprocessing, the unused fissionable material can be recovered and waste products can be removed. One of the products formed during reprocessing is the famous element plutonium.

6
PLUTONIUM: AN UNNATURAL ELEMENT

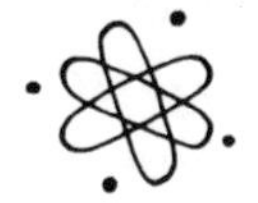

Practically no plutonium appears in nature. Very small amounts of plutonium result from the interaction of cosmic-ray neutrons with uranium. While a uranium deposit in the southeastern part of Gabon was being mined, it was discovered that some plutonium existed in a natural fission reaction that started two billion years ago in West Africa. It had long been dormant, but finding the remains of this unusual fission reactor was most exciting. Obviously, it is not a source of plutonium today.

Uranium-235 releases heat, two lighter atoms, and either two or three neutrons when it splits in today's reactors. The free neutrons may enter a nucleus of another uranium-235 atom; they may be captured by part of the reactor structure; they may escape from the system; or they may collide with atoms of uranium-238, the more plentiful kind of uranium present in the fuel. When the latter happens, the uranium-238 may fission, but since it is relatively stable, it will probably absorb the neutron to become uranium-239. This radioactive substance decays, and through a series of steps becomes plutonium-239. Similar radioactive decay processes take place in other atoms in reactor fuel to form other types of plutonium. Some plutonium atoms fission, releasing radiation and new elements. Two forms of plutonium are active in providing the heat in a reactor fueled with enriched uranium. Near the end of the fuel's useful life, plutonium-239 and plutonium-241 contribute about as much energy as uranium does. The plutonium formed in uranium-fueled reactors is man-made fuel. Not all the uranium-235 and plutonium in the fuel rods can be used to produce power before the rods must be removed. Recycling the fuel is a subject of much controversy because plutonium, as well as being especially valuable, is especially dangerous.

Most of the plutonium in spent fuel rods can be reclaimed. A small amount decays during storage, and a small amount is left even after reprocessing. Plutonium reclaimed through reprocessing can be used in a number of ways. Batteries powered by plutonium are considered suitable for powering satellites, remote weather stations, navigational beacons, and cardiac pacemakers implanted in patients with heart blocks or irregular heartbeats. These pacemakers are used to set the heart rate when a patient's natural control by the nervous system is no longer effective.

The makers of nuclear-powered pacemakers insist that the plutonium they contain could never join the nuclear wastes that might pollute the environment. Some physicists disagree, and are concerned. There is a very small amount of plutonium in every nuclear-powered heart pacemaker. Suppose a person who has been fitted with one dies and is cremated. According to the manufacturers of the pacemakers, crematory tests show that the plutonium, which is incorporated into a hard ceramic pellet, does not break down into particles. The manufacturers say that the only way plutonium could be released from a pacemaker would be if someone intentionally cut a pacemaker open. But many other people raise the question of plutonium possibly being released from pacemakers during fires where temperatures are high enough to melt the pacemakers and release plutonium.

Since the Russian satellite Cosmos 954 crashed

in the Canadian wilderness in January 1978, there has been concern about the possibility of the release of plutonium from a nuclear-powered satellite that misfunctions and falls back to earth. Cosmos 954, which fell far from populated areas, released some radioactivity from its uranium-powered reactor. Although the amount of plutonium in satellites is small compared with the uranium reactors in Cosmos 954, release of plutonium must not happen at all.

SNAP is a System for Nuclear Auxiliary Power that is used in weather satellites. The power comes from the decay of plutonium-238, an excellent source of power for a long period of time. The plutonium is well protected, and in one case where a satellite malfunctioned, the plutonium power source was recovered intact with no apparent leakage. In another accident, the generator fell into the Pacific Ocean where it was not recovered; the protective graphite container may remain intact for many years. Plutonium-238 has a half-life of 86.4 years and is considered dangerous for much longer than that.

The power source of a SNAP experimental satellite vaporized in another accident when the satellite failed to achieve orbit and burned during reentry into the atmosphere. The amount of plutonium released in this accident in 1964 is estimated to have been greater than the plutonium-238 fallout from bomb testing of earlier years.

In 1978 the loss of a plutonium-powered spy device in the Himalayas 12 years earlier was revealed to the public. This nuclear power pack was reported to contain 2 to 3 pounds of plutonium-238, which was doubly insulated by encapsulation in leaktight capsules. There was concern about possible pollution of the sacred Ganges River, which has its source in this mountain range. No trace of the device has been found.

Before wastes can be reprocessed, they must be shipped to a reprocessing plant from the storage pool at the reactor site. Spent fuel rods must be transferred to massive, heavily shielded and cooled portable lead casks. The casks used for shipments by rail or truck have been called the most carefully designed shipping containers ever made. Before January 1976, when the New York City Council banned the shipment of nuclear wastes through the city, six to eight shipments were made twice each year: the New York City police escorted a truck that carried the spent fuel rods from Brookhaven National Laboratories in Long Island through New York City to New Jersey. From New Jersey to the reprocessing plant in South Carolina, the truck went on its way just as any other truck would. But though there was never an accident during 25 years of shipments, the prospect of one was so alarming that the Department of Health's Bureau of Radiation Control was instrumental in effecting the ban on further shipments after 1976. It was estimated that the release of 1% of the cargo of a truck in a densely populated part of New York City would cause 100,000 deaths in a short time after an accident, and that as many as a million people might develop cancer from exposure to radiation. Since casks for carrying high-level radioactive materials have been subjected to torture tests, which include jet fuel fires that completely engulf the cask, the likelihood of an accident during shipment of high-level wastes is considered remote. If a cask is involved in a truck or train accident, the damage is expected to be superficial. In spite of this assurance, other communities beside New York City have voted against the shipping of nuclear wastes on their roads.

Estimates about the danger from exposure to plutonium vary and are controversial. According to one estimate, if 0.001% of the accumulated plutonium production between 1977 and 2000 escaped to the environmen 160 cases of lung cancer would develop; another estimate places the number of cases of lung cancer at 116,000. Part of this variation is due to a disagreement about the way plutonium affects the human body. Most forms of plutonium (atomic weights of 238, 239, 240, and 242, but not 241) emit alpha radiation, the kind that can be stopped by a sheet of paper. Plutonium-242 decays to produce another alpha emitter, americium-241. Since alpha radiation has weak penetrating power, one may wonder why there is such a fear of plutonium. But it is this property that appears to make plutonium especially dangerous. If a speck of plutonium is breathed into the lungs, it affects only those cells that are very close to it, but it keeps affecting them strongly. A particle that penetrates deep into the lung may remain there for a year or more.

The difference of opinion about the potential effect of breathing a virus-size speck of plutonium-239 is a striking example of the controversies about the dangers of nuclear radiation. Some authorities consider plutonium mainly as a treasure to supply large amounts of reactor fuel in a world that is increasingly hungry for energy; others warn that increased production and use of plutonium can be a serious health hazard. The controversy about a speck in the lungs is often called the "hot particle" argument. Both sides agree that a speck can expose surrounding lung tissue to intense radiation. Some experts say the continuing radiation is wasted on nearby cells that are already dead and that the overkill cannot be considered an absorbed dose. Others believe that an individual particle can deliver larger doses of radiation to tissue in the lung and that the plutonium may eventually be carried to lymph nodes surrounding the lungs, the liver, and the bones, where it can cause further damage.

The "hot particle" theory, first presented in 1950, has been vigorously supported by Drs. Arthur Tamplin and Thomas Cochran. It has been just as vigorously opposed by a committee of the British Medical Research Council and by a group of other Americans. While some studies discredit it, the "hot particle" hypothesis cannot be said to have been completely discredited.

There may be plutonium in the lungs of every person on earth as a result of the fallout from bomb testing many years ago. Those who feel that its potential for hazard has been greatly overstated claim that no human deaths have been caused by cancer from plutonium. Others question this, since the results of exposure to plutonium may not be obvious until many years afterward. Cancers may not develop for decades.

Since authorities disagree, the U.S. Transuranium Registry in Richland, Washington, is collecting information on workers who have been exposed to plutonium. Uptake, distribution, and retention of plutonium and other transuranium elements (those with atomic numbers greater than that of uranium) are being studied. Some of the 5800 workers identified in this program have given permission for autopsies, but the study is still young and the number of people involved is still small. Much remains to be learned.

What happens when small amounts of plutonium escape from the nuclear cycle? As a result of accidents at the Rocky Flats, Colorado, plutonium fabrication plant for the weapons program, a substantial amount of plutonium was released into the environment by fires in 1957 and 1969. More plutonium escaped over a long period from leaky drums of plutonium-contaminated oil. In this semi-arid region, escaped plutonium was spread primarily by the wind. According to one report, the plutonium concentration in the soil at the east side of the plant is 1300 times that from fallout and it is largely concentrated in a thin surface layer of soil. A more recent study gives a lower value to the plutonium contamination outside the Rocky Flats plant. There is agreement, however, that the plutonium is spread from one place to another largely by wind that carries dust particles containing plutonium. The particles appear to be lifted from surface soil again and again, making it possible for people who live miles away to inhale particles before they resettle. The northwest suburbs of Denver are 6 to 8 miles from the plant.

A different pattern has been followed by plutonium that was part of the low-level radioactive waste stored at Maxey Flats, Kentucky. Between 1963 and 1972, about 2.5 million cubic feet (70,000 cubic meters) of radioactive wastes were buried in trenches at the site, located atop a dissected plateau near Morehead. The upland is 250 to 350 feet (75 to 105 meters) above the surrounding valleys, which contain small creeks. The solid low-level wastes, which contained some plutonium, were buried in trenches. Measurements have shown that substantial amounts of plutonium migrated hundreds of yards through the soil, probably with the help of groundwater movement. Following the many reports that have been released on data collected at Maxey Flats, it has been recommended that a buffer zone of several thousand feet to several miles be established around such low-level waste disposal sites.

There is also controversy about the many reports that have been made on the effects of plutonium on entire living organisms. Since plutonium is not soluble in water, it was believed that small amounts did not affect plant life. However, laboratory studies in the last few years have disproved some of the original assumptions, revealing that plants absorb plutonium from the soil just as they absorb other trace elements in very small amounts. Early scientists may have been misled by inferior measuring techniques. Drs. Raymond R. Wilding and Thomas R. Garland of Pacific Northwest Laboratories, using techniques that involved very precise measurements, found concentrations of plutonium in the root systems of plants grown in soil containing the element. Many root plants (such as beets, carrots, and potatoes) are eaten by humans. It is important to know if plutonium from leaking waste has reached this food, even though eating plutonium appears far less hazardous than breathing it.

Whether you consider plutonium to be trash or treasure, it is important to keep it contained. Spent fuel rods that contain large amounts of potential fuel are being held for reprocessing in many storage pools at nuclear reactor sites. Will these storage pools become plutonium mines for terrorists? Or can this kind of trash be safely recycled into treasure?

7
NUCLEAR WASTES: THROWAWAY OR RECYCLE?

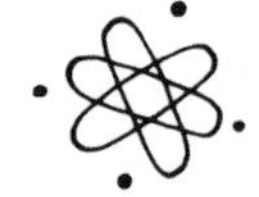

The controversy is brisk between people who believe in the throwaway fuel cycle and those who favor recycling spent fuel for further use. It is agreed that plutonium is a substance that must be respected and guarded for thousands of years, and there is fear of the problem of keeping plutonium from terrorists and from countries that do not already have nuclear power. The enriched uranium used to power conventional reactors contains 2 to 4% uranium-235 and 96 to 98% uranium-238. This fuel-grade uranium can be made into weapon-grade uranium only in an enrichment plant. Plutonium can be used as a weapon more easily than uranium, with a conventional, though complicated chemical process.

No one knows how to make plutonium unusable in weapons.

Concern about plutonium prompted the famous decision of April 7, 1977, announced by President Carter: "We will defer indefinitely the commercial reprocessing and recycling of the plutonium produced in the United States nuclear power programs. From our own experience we have concluded that a viable and economic nuclear power program can be sustained without such reprocessing and recycling. We will redirect funding of the United States nuclear research and development program to accelerate our research into alternative nuclear fuel cycles which do not involve direct access to materials usable in nuclear weapons."

This holding pattern on U.S. nuclear fuels began in 1977, and already as many as 18 countries, besides those that have exploded nuclear weapons, are believed to have access to reprocessing technology. An International Nuclear Fuel Cycle Evaluation began in 1977, to extend over a two-year period during which "proliferation-resistant" alternative fuel cycles could be assessed. The Barnwell nuclear fuel plant at Barnwell, South Carolina, built to accommodate the fuel recovery needs of fifty to sixty large, water-cooled, uranium-fueled reactors, could not be licensed without reversing the policy of halting commercial reprocessing of spent fuel. In 1978 the Department of Energy signed a contract for nonproliferation research and development at the Barnwell plant. Since this newly constructed plant was not contaminated with radioactivity, experimental processes could be evaluated. The plant was designed for techniques that have been used for some 30 years in about a dozen fuel-reprocessing plants throughout the world. However, with minor modifications it could be used to experiment with processes considered less vulnerable.

In the technique known as Purex, spent fuel rods are transported in casks from the storage pools at nuclear reactors to reprocessing plants. The rods are removed from the casks under water and stored in pools much like those back at the reactor site until they lose some of their shortlived radioactivity.

During reprocessing the rods are sheared into many short pieces inside a concrete-walled work area known as a hot cell. Materials are handled by remote control since there is still considerable radioactivity. Rod segments are dropped into nitric acid to dissolve the exposed fuel pellets. Through chemical separation processes almost all the uranium and plutonium are recovered and can be refabricated for use as a new fuel. A waste management program must deal with the remaining intensely radioactive mixture.

The uranium and plutonium made available for fuel, and the highly radioactive residue make reprocessing an international problem. While the wastes are being recycled to make the spent fuel usable, the plutonium is more accessible to terrorists. Even recovered uranium may appeal to countries that want only a few bombs for diplomatic leverage or nuclear blackmail.

Several individuals have shown that they can draw plans for homemade nuclear bombs with knowledge that is available to the public. A 22-year-old former Harvard student who had studied college-level physics for only a year drew designs for a series of nuclear weapons. A weapon designer described the designs as the most extensive and detailed he had seen outside the classified literature. Students at Massachusetts Institute of Technology and at Princeton University have developed plans on their own, too. This has helped alert government officials to the need for increased safeguarding of nuclear materials.

The designs by students are considered somewhat primitive, but it is possible that professional bomb makers or qualified scientists may change their loyalty and contribute expert help in producing a bootleg bomb. For weapons, reprocessed plutonium is better than reactor uranium. The combination of different kinds of plutonium reclaimed from spent fuel rods is called reactor-grade plutonium: it contains more plutonium-240 and plutonium-241 than the plutonium used in making weapons, and this makes an explosion weaker and less predictable. However, a bomb made from reactor-grade plutonium would be devastating compared to non-nuclear bombs. Also, a country might produce weapons-grade plutonium in its own reactors. That weapons-grade plutonium can be made from a smaller amount of undesirable kinds of plutonium by tampering with a reactor has already been demonstrated in India.

Security measures for reprocessed fuel may have to be as stringent as those used to protect plutonium used for weapons. Some scientists have suggested this as the way to safeguard plutonium in reprocessing.*

Estimates vary about how much reprocessed plutonium is needed to make a bomb or other nuclear weapon. Even if the amount is too large to make bootleg bombs likely, reprocessed fuel must be carefully guarded from terrorists for another reason, too. If a small amount of plutonium were dispersed through the air-conditioning system of a large building or released in a highly populated area of a large city,

*Jan Prawitz, "Is Nuclear Power Compatible With Peace?" in *Facing Up to Nuclear Power*, Philadelphia: Westminster Press, 1976, pp. 103–104.

what would happen? No one really knows, but it has been calculated that 15 grams of powdered plutonium placed in the air-circulation system of a large office building might result in only one plutonium death because only a small percentage of the plutonium would be inhaled. This does not include deaths from cancer, which might appear many years later. It has been noted that such predictions might discourage the use of plutonium as a terror weapon, however, since other methods of destruction are more easily available and more dramatic.

In the throwaway approach, where fuel goes through the system just once, there are some built-in safeguards. The fissionable material in the fuel is diluted (even enriched uranium is mostly uranium-238) and would need much more enrichment before it could be used as a bomb. Another is the intense radioactivity of the spent fuel rods. But what if these fuel rods remain in storage pools at reactor sites until the intense radioactivity diminishes? Radioactive iodine-131 has a half-life of 8.14 days. Some other radioactive materials in the spent fuel rods have short half-lives, too. The overall radioactivity and the heat decrease to 1/10,000 during several months of storage. According to Dr. Chauncey Starr, president of the Electric Power Research Institute, the used fuel has lost enough of its radioactivity to be vulnerable to theft after ten years. Storing fuel elements indefinitely creates plutonium "mines" because the availability of plutonium increases as the number of older fuel rods grows in the storage pools. According to those who favor reprocessing, plutonium is less available if it is used again as fuel rather than left to accumulate in storage areas.

Nearly all the reprocessing plants built by the government for military programs in the last 30 years have used the Purex process. This is an extraction process that uses a chemical solvent to separate plutonium and uranium from the fission products in spent fuel.

Early in 1978 the development of a recycling system to prevent the diversion of plutonium was announced by research officials in the United States and Great Britain. Known as Civex, the process avoids purifying weapons-grade material at any stage. Reprocessing plants that use the Civex system will carry out all steps in hot cells by remote control. The process, which combines known techniques, has been tested on a laboratory scale. It begins the same way as Purex, with spent fuel rods being dissolved in acid. But pure plutonium and uranium are not separated from the waste fission products: they are contaminated with very toxic radioactive elements such as zirconium, ruthenium, and niobium. The Civex method cannot be used on a large scale, however, until it has been explored more thoroughly for some years.

One reason for the great interest in reprocessing is the possibility of using recycled fuel in breeder reactors. Theoretically, breeder reactors can create four units of plutonium for every three that they use. In a fuel-hungry world, this sounds almost too good to be true. Certainly breeding is not a magic wand; not all breeder reactors are welcomed by everyone.

Two kinds of breeder reactors have been explored experimentally. The light-water breeder reactor, sometimes referred to as the slow breeder because it uses neutrons at low speed, is an alternative to conventional nuclear power plants and to the controversial fast breeder reactor, which will be discussed later. The slow breeder uses a cycle based on the element thorium, which is abundant and inexpensive. It is not without problems and risks; one difficulty is in obtaining uranium-233, a form of uranium that must be mixed with the thorium, and which is not found in nature. The original supply must be made in reactors from uranium-235. Once enough uranium-233 has been bred, the light-water breeder does not need any more natural uranium-235.

President Carter dramatically ordered the

experimental light-water breeder program at Shippingport, Pennsylvania, to proceed to full capacity by writing instructions on an electronic blackboard in the White House. The fuel cycle at Shippingport begins with uranium-235. Fuel pellets of fissionable uranium-235 produce neutrons that are absorbed by thorium-232. When a thorium atom absorbs a neutron, it is converted into uranium-233. Uranium-233 fissions, and more uranium-233 is produced from the thorium-232. This thorium-232 in rods surrounding the core is known as a blanket. It sponges up neutrons, much like the control rods in conventional reactors, and makes fuel (uranium-233) in the process.

Less plutonium is produced in slow breeders than in other reactors that produce plutonium. Unfortunately, uranium-233 may be as useful for nuclear explosive devices as plutonium. However it is somewhat more difficult to handle because a small amount of uranium-232 is produced with it. This uranium-232 produces gamma rays, which penetrate deeper than the alpha rays emitted by plutonium.

Uranium-233 can be diluted with uranium-238, making it useless for nuclear explosions, but this puts a plutonium producer back in the reactor. A thorium-uranium cycle will increase safeguards against diversion or theft only if the slow breeder is designed to minimize the production of plutonium and the occurrence of high concentrations of uranium-233. A very delicate balance is needed and this is difficult to achieve.

Since the slow breeder reactor makes slightly more fissionable material than is put into it, it is considered a possible way to stretch fuel supplies for many years. The radioactive waste total is estimated to be the same as that from conventional reactors. There is still a need for reprocessing and refabrication with the slow breeder, and high-level radioactive wastes are an unwanted by-product of nuclear fuel reprocessing. There are also the problems of keeping a good balance in this closed-cycle system. While tests of breeding in light water have begun, not all the questions have been answered.

The fast breeder reactor is considered the best hope by many people who look to recycling plutonium. Fast breeder reactors have been operating experimentally for a long time. The first reactor to generate electricity was an experimental breeder, in 1952.* Experiments are expected to continue for a long time before this kind of reactor is considered ready to produce commercial electricity. Although it is a major part of the nuclear cycle if the throwaway system

*WASH 1214, Vol. II, U.S. Atomic Energy Commission, 1972.

is not chosen, many people feel that the breeder reactor will never be safe. Meanwhile, France, Japan, and other countries are moving ahead with fast-breeder technology.

Why is the breeder so controversial? Here is the basic way a breeder works. Actually, a fast breeder does not breed fast, but uses fast neutrons to breed slowly. Fuel rods that contain plutonium-239, uranium-235, or a mixture of them, plus uranium-238 are placed in the core. The area known as a blanket, which surrounds the core, is made of rods that contain, most often, uranium-238 from other reactors. When neutrons are captured by uranium-238 in a fast breeder reactor, the eventual result is plutonium-239. Not all neutrons that enter the blanket are captured by the uranium-238; a fairly large proportion are reflected back to the reactor core where they are useful in increasing fission there. But more uranium-238 absorbs neutrons in a breeder reactor than in a reactor where temperatures are lower. The high temperatures in a breeder tend to increase the range of neutron speeds and give more chances for neutrons to travel at the speed needed for absorption by uranium-238. At intervals the blanket is removed and sent to a reprocessing plant where plutonium is extracted and used to make new fuel rods for another reactor.

A breeder does not really make more fuel than it uses, but it does change fertile uranium-238 into plutonium-239, which fissions, and it does so faster than it consumes the fuel that is fissioning. Technically the uranium-238 is a fuel.

Because of the very high temperatures and neutrons traveling at higher speeds than in conventional reactors, the cooling substance must be something other than water, which would have to be kept under great pressure—and besides, slows the speed of neutrons. The most popular coolant in fast breeder reactors is liquid sodium, which is so reactive that it is explosive when it comes in contact with water. It becomes intensely radioactive, so radiation shielding must be used to protect the workers who are near the sodium that has circulated through the core and blanket to the area where the sodium gives up some of its heat to water in a heat exchange system.

Although the liquid metal fast breeder reactor (LMFBR) presents some new safety problems, it reduces some inherent in conventional reactors, and many people feel it is needed to supply the energy required in future years. Proponents claim that energy from breeder reactors may be 72 times as great as that from conventional reactors. Opponents claim that the ratio may be closer to 5, over a period of a hundred years.

Estimates of the number of years before a breeder produces twice the amount of fissionable material it uses vary from eight to sixty years. Opponents observe that if sixty years is correct, this could be longer than the life of the plant. No one really knows the answer yet.

To realize the benefits of the fast breeder reactor, spent fuel elements from other reactors must be reprocessed after being transported from storage pools. They must be fabricated into fast reactor fuels, and transported to the fast breeder reactors. Plutonium must be recovered from the blanket at intervals after operation of the reactor begins. If fears that plutonium can be diverted to weapons are quelled by never separating this element from a high level of radioactivity, can other problems be solved? Even if a research program is vigorously pursued, fears about diversion and safety may persist for a decade.

Suppose breeder designs are perfected to the point where safety is no longer a problem. And suppose the recycling process eliminates the risk of channeling plutonium into military weapons. There would be less volume of waste to store, if fuel is recycled. Much of the radioactive content of fuel rods would be turned into a new fuel supply, but not all of it would be consumed.

The moratorium on reprocessing delayed the

decision about recycling nuclear fuel. It slowed progress, which many scientists felt was too fast. But while it gave time for more research in the United States, other countries moved forward. In the spring of 1978 the British House of Commons voted to build a reprocessing plant at Windscale in Northumberland. The full-scale commercial facility will have the capacity to reprocess 1200 metric tons of spent fuel each year. Some of this will come from countries far away; for example, the United States agreed to the British reprocessing of 42 bundles of fuel from a nuclear power station in Japan. The Japanese station was running out of storage space. Permission from the United States was needed in this case because the fuel was originally of United States origin.

Spent fuel rods from countries far from France have been arriving on the French coast awaiting reprocessing at Cap de la Hague, near Cherbourg. There the government plans to add two more reprocessing plants to its military plant, which has been modified for commercial use. Many French citizens are protesting the government program, claiming that it will turn France into an international trash heap. France plans to dispose of 2400 tons of spent fuel each year when capacity is reached. Low-level radioactive wastes and liquid wastes will be buried in containers underground or at sea. Plutonium which comes from reprocessing other wastes will belong to the countries that sent the spent fuel to France; returning it to them will involve international treaties.

In October of 1977 President Carter announced that the U.S. government would take spent reactor fuel from a utility company's storage pools for a fee and assume complete responsibility for the fuel's safe handling and storage. In this throwaway nuclear cycle, utilities receive no payment for the plutonium in spent fuel rods.

When the announcement to store spent fuel was made, the Department of Energy had no schedule of charges for its handling and disposal, and no firm policy for disposal of nuclear wastes had been established. Large amounts of wastes can accumulate in a throwaway system. With reprocessing and recycling, some of the problems of disposal of plutonium may be eliminated—this would depend on many factors.

No matter which path is taken, there are many nuclear wastes to store. What should be done with these radioactive atoms?

8
THE VARIETY OF NUCLEAR WASTES

The "atomic trash basket" contains a variety of wastes. From the beginning of the nuclear cycle to the end, even if recycling is the method of choice, there are places where unwanted materials appear. These materials, which cannot be used, create one of the major problems in nuclear energy.

We have mentioned mill tailings under buildings as a contamination problem. Radioactive wastes can seep into streams and follow several paths to people. Water in pastureland is absorbed in plants and by cattle. Wastes can become part of plants that live in water, and food for insects and fish. Plants, fish, and insects may become food for people directly or indirectly as part of a complex food chain. Some radioactive trash reaches people directly as drinking water or indirectly in plants and animals used for food. The radioactive atoms in trash remain hazardous for so long that great care must be taken to control the number of such atoms that enter water.

Wastes from preparation of fuel and fission products from chain reactions in reactors must be isolated from the environment. The front end of the nuclear cycle (mining, milling, conversion to uranium compounds used for fuel, enrichment of fuel, fabrication of fuel) does not produce wastes as radioactive as those produced by fissioning of fuel in the reactor. The fragments of fissioning uranium nuclei include hundreds of chemicals, some of them extremely toxic to public health.

Compared with the highly radioactive (high-level) wastes of such fuel, wastes from the beginning of the cycle and materials that have been exposed are classified as low-level wastes. Much low-level trash is similar to common trash except for its contamination by radioactive atoms. Millions of cubic feet of radioactive tools, wiping rags, glass, gloves, worn-out machinery, and general garbage are stored in nuclear burial grounds. Some low-level wastes are unusual; for example, the carcasses of animals that have been treated with radioactive substances.

Much low-level waste has been compressed by a compactor to force the material into a disposable container, baled by a compressor, or bagged by a machine that compresses waste into a predetermined shape for convenient disposal. Incineration has reduced the volume of wastes, especially in countries where land is scarce.

According to a report by the National Research Council's committee on radioactive waste management, the volume of low-level solid waste produced in 1976 at the Energy Research and Development Administration laboratories and plants was about the same as the solid wastes produced each year by a town with a population of 55,000. Commercial nuclear plants produce about the same amount annually but it is expected to increase dramatically.

Besides low-level wastes produced by nuclear research laboratories and power plants, there is the escalating problem of nuclear plants that are no longer usable. What can be done with a decommissioned, radioactive plant? Several scenarios have been suggested.

1. *Mothballing* means guarding the plant from access by the public day and night until enough time lapses to make it safe from radiation hazards. This could be many centuries.

2. *Entombing* is accomplished by welding shut the entrances to the plant, sealing them with concrete, and installing alarm systems to alert securi-

ty guards to any attempts at breaking and entering. This too would be very long-term.

3. *Mothballing for 100 years and then dismantling* without remote control would make dismantling less expensive because the plant would be less radioactive after a century.

4. *Entombment combined with dismantling after 100 years* is preferable to the third scenario, unless another nuclear power plant is nearby so that security guards familiar with the decommissioned plant can serve as monitors to both.

Imagine the problems of dismantling a solidly built structure that is too radioactive to touch! The workmen have to use remote control torches and watch closed-circuit television to see what they are doing. They slice up the reactor bit by bit, hoisting slabs by crane into an 8000-gallon water tank which shields them from radioactivity. Eventually the pieces are enclosed in sealed containers for transportation to a burial site far away. Estimates for dismantling the average commercial reactor are $100 million and more. After 100 years, the decay of cobalt-60 permits dismantling with care but without the need for remote control. Remote manipulation methods would not be needed at a plant dismantled after 100 years, which would cut costs by half. If a plant is never to be dismantled, permanent mothballing will be needed for complete safety because of the presence of long-lived nickel-59 and carbon-14. In this case "permanent" means half a million years.

Suppose a nuclear power plant can be built so that shielding, walls of the building, and other parts of the plant can remain after the 40-year useful life, and that internal operational parts can be replaced. The panel on land burial of the committee on radioactive waste management of the National Academy of Sciences suggests that much more thought be given to the design of the basic structure of these buildings. Thus commercial power-generating reactors of the future would have their useful life extended. The problem of dismantling and decommissioning nuclear power reactors and other nuclear facilities is growing, and may loom larger in the future.

The nuclear burial grounds of low-level wastes include burial in trenches, which is now restricted to waste that produces radioactivity in the same range as that produced by radium in the earth's crust. Since some low-level wastes contain long-lived radioactive materials, many investigators are calling for new studies on the actual rate of migration and the speed with which such wastes are released into the environment. Releases to date are not considered a major health problem, but there are indications that not everything has worked as planned. In some burial sites that were intended to contain radioactive materials for hundreds of years, migrations have been detected far beyond expectations in a dozen years. The leakage of low-level wastes at Maxey Flats, Kentucky, has already been mentioned.

A report to Congress, "West Valley and the Nuclear Waste Dilemma," notes that the first shipment of low-level radioactive wastes was buried at a 6-acre site in New York in November 1963. Trenches were dug 30 feet wide, 20 feet deep, and up to 800 feet long. Wastes were put in them and earthen caps placed over the trenches when they were filled. The trenches began to fill with water in the mid-sixties. In the early 1970s the New York State Department of Environmental Conservation detected radioactivity at small levels in streams adjacent to the burial site. Further studies showed extensive water infiltration in a number of trenches which transported radioactive materials to a nearby stream, plus considerable surface erosion over some trenches. Water was pumped out of the trenches, burial operations were stopped, and water was treated before being pumped away from the site. Pumping continued in 1977 and there appeared to be no significant health hazards from the wastes, but continued remedial action was recommended.

Low-level wastes are less troublesome than the

water which carries away some of the heat generated by fission. (Only part of the heat is used to produce the electricity, so excess heat must be removed.) Waste water is not radioactive but it is hot. Pouring back into rivers and lakes, it affects the plant and animal life. Large cooling towers using a combination of air and water, or cooling lakes or ponds, are used to prevent damage.

With careful controls, thermal wastes from the nuclear process may even be put to use. For example, at Vermont Yankee Power and elsewhere, efforts are being made to develop commercially profitable greenhouses. At the Vermont Yankee plant, an aquaculture laboratory is dedicated to the Atlantic salmon restoration program. It seems possible to raise fingerling salmon to a size where they are ready to molt by using water from the cooling system at a temperature of 60° F(16° C). The growth rate is doubled or tripled by the warm water. Experiments are being carried out to enhance the growth of plants, too. Each of four greenhouses is heated in a different way so that cost comparisons can be made. A unique feature in this experiment is the use of a special machine to digest manure from an adjacent dairy farm. Waste heat from the nuclear power plant is used to preheat the manure mixture (slurry), before introducing it into the main digester. Sufficient manure may be available from this source to adequately heat one of the experimental greenhouses solely by burning biogas (from the manure).

Since as much as 50% of total energy input is lost in conversion of heat to electricity in large nuclear and fossil-fuel electrical generating stations, controlling waste heat is important. While much of the waste heat in a fossil-fuel plant goes up the smokestacks, waste heat from nuclear plants enters the environment through the water. If the experiments with using such wastes to increase food supply work out, the wastes may even be an asset.

High-level wastes from nuclear facilities are the part of the nuclear trashbasket that causes the most controversy and concern. The site at West Valley, New York, mentioned earlier in connection with low-level wastes, has become famous for problems of high-level wastes. This was the first commercial reprocessing plant in this country. Reprocessing began at West Valley in 1966 and ended six years later, when West Valley was closed for renovation and expansion. The original owners apparently did not recover their initial capital investment, and their operation was not trouble-free. There were problems in staying below federal limits on radioactive discharge and employee exposure to radioactivity. Soon after the shutdown, ownership changed. The new owner later announced that the plant would not reopen, but unfortunately this was not the end of the story.

The dilemma at West Valley is probably unique, the wastes left there and the way they are stored differing from what is likely to be found in any future fuel reprocessing facility. The West Valley problem has been compared with the vastly larger problems at the Hanford Reservation near Richland, Washington, and at Savannah River, South Carolina, where fuel reprocessing continues to be carried out in connection with materials for the United States military program. A characteristic of waste management at all three sites is the precipitation of insoluble, highly radioactive sludge which can only be redissolved by making it acid, a process that would also dissolve the carbon steel tank walls.

Since the carbon steel tanks at West Valley cannot last indefinitely, the waste management program was a temporary one. Unfortunately, the following trash is still there:

- a burial ground for solid wastes with an intermediate level of radioactivity;
- a 750,000-gallon carbon steel tank holding approximately 600,000 gallons of nuclear wastes which have been neutralized and

sludge with a high level of radioactivity;
- an empty 750,000-gallon carbon steel tank;
- a 15,000-gallon stainless steel tank storing approximately 12,000 gallons of acid, high-level wastes cooled to minimize damage due to hot acids; and an empty 15,000-gallon stainless steel tank.

The shipment of high-level radioactive wastes, so that the liquid wastes could be pumped out and the contents transferred to another site, is not permitted at present. The wastes in the carbon steel tank present an even greater problem. They were neutralized to prevent corrosion of the tank, but if they are made acid again to chemically dissolve the sludge, the acid will attack the tanks. Removal by mechanical means is complicated by the presence of obstructions inside the tank bottoms and by limited access to the tanks. What is to be done at West Valley, and whose responsibility is it? These are questions which have been asked again and again.

New federal regulations forbid storage of high-level liquid wastes; they must be solidified within five years after they are produced and sent to a federal repository. So the situation at West Valley cannot happen again.

According to Dr. Gene I. Rochlin, a physicist and research specialist in the Institute of Government Studies, University of California at Berkeley, there are lessons to be learned from West Valley. He suggests that mechanisms are needed to provide for assuring legal and financial responsibility for a very long time.*

The poisoning of fish and plants from the hot-water waste at some nuclear power plants has received more publicity than some other wastes in the nuclear trashbasket. Still, knowledgeable people are concerned about the safe storage of wastes which contain radioactive materials such as the following man-made transuranic elements:

	Half-life (Years)
Neptunium-237	2,140,000
Plutonium-238	86
Plutonium-239	24,390
Plutonium-240	6,580
Plutonium-242	379,000
Americium-241	458
Americium-243	7,950
Curium-245	9,300
Curium-246	5,500

*Gene I. Rochlin, Margery Held, and Barbara Kaplan, "West Valley: Remnant of the AEC," *Bulletin of the Atomic Scientists*, January 1978, p. 24.

It has been suggested that a radioactive substance be considered safe after the passage of 20 half-lives. At this standard, plutonium would have to be isolated from the environment of plants and animals for about 500,000 years.

One problem of safe disposal of low-level wastes is that some of them include small amounts of radioactive materials with long half-lives. Long-lived transuranics in low-level wastes may have to be exhumed and isolated with the high-level wastes.

Large amounts of wastes have been accumulating for decades from military and research and development efforts, as well as from commercial nuclear power plant operations. While the military wastes have a lower per unit volume of radioactivity than commercial wastes, they take up much more space. It is estimated that by the year 2000 two repositories will be needed for military wastes for every one-and-a-half repositories for civilian wastes. Just how much is in the nuclear trashbasket now?

In the fall of 1977 the Comptroller General sent a report titled "Nuclear Energy's Dilemma: Disposing of Hazardous Radioactive Waste Safely." This stated that about 71 million gallons of high-level waste produced by the Energy Research and Development Administration's plants have been "temporarily" stored in steel tanks at the Hanford facility in Richland, Washington, and at the Savannah River facility in Aiken, South Carolina. About 3 million gallons are stored in underground tanks and bins at the Idaho National Engineering Laboratory at Idaho Falls, Idaho.

About 13 million cubic feet of transuranic contaminated waste from military and research activities have been buried or stored retrievably at five principal shallow-land burial sites of the ERDA. This waste is contaminated with about 1,000 kilograms of plutonium. Some of the 1.3 million cubic feet of radioactive waste generated by the ERDA each year is contaminated with transuranic elements. About 7 million cubic feet of commercial transuranic contaminated waste are expected to accumulate by the year 2000.

About 600,000 gallons of high-level waste have been generated from commercial reprocessing activities and are currently stored at West Valley, New York. Should commercial reprocessing operations resume, estimates are that by 2000 an additional 152 million gallons of high-level waste will be generated.

The ERDA estimates that 1988 is the earliest that a geological waste disposal facility or other storage facility to receive spent fuel can be ready. By this time the nuclear industry may have a severe shortage of storage capacity.

Spent fuel and transuranic contaminated waste can be as hazardous to the public health and safety as high-level waste. While the federal government has extensive regulatory and public oversight over most nuclear plant operations because they use nuclear materials, the same degree of public protection and independent oversight is not required for the storage and/or disposal of nuclear materials. This needs to be changed.

Information for one 60-page review was obtained by examining environmental reports. geologists' reports, correspondence and other documentation, and interviews of officials at commercial power plants, the ERDA, the National Regulatory Commission, and so on.

Many reviews of the problem of nuclear wastes are in progress and not everyone agrees about what is in the nuclear trashbasket or what dangers may come from future wastes. Considering all sources of nuclear wastes, the greatest exposure of the population may come from gases that are released into the atmosphere. Radioactive krypton and tritium, a form of hydrogen, are gaseous wastes from reprocessing. Krypton-85 is easier to collect than tritium (hydrogen-3). The difficulty of collecting tritium may become a limiting factor in the production of nuclear energy. Certainly, disposal of these gases through stacks which send them into the atmosphere is unacceptable for any period of time. Since radioactive gases receive far less publicity than other radioactive wastes, many people are unaware of the problem.

Wastes containing radioactive materials may be liquids, solids, or gases. They give off various levels of intensity for various lengths of time. There is no way of hurrying the decay from radioactive to nonradioactive; final disposal must be by natural decay. Processing, packaging, shipping, and storage must all be aimed toward leaving wastes where they will not contaminate the environment of today's and tomorrow's generations.

9
SHOOT IT INTO SPACE, BURY IT, PACK IT IN ICE?

The U.S. government has announced a plan to assume ownership of spent fuel for a onetime fee paid by nuclear facilities, but no state wants a temporary or permanent federal repository in its "backyard." A federal repository is a facility operated and owned by the government for storage or disposal of radioactive wastes, which may come from ERDA sites and/or from the nuclear industry. Such a facility may be used for extended storage of commercial spent fuel, permitting a decision on whether the fuel will be reprocessed or discarded.

The search for places to store high-level radioactive wastes is not new. As long ago as 1957 permanent disposal was recommended by a special committee of the National Academy of

Science–National Research Council. Since then many ideas have been explored. A well-known one is to shoot long-lived wastes into space via rocket. This idea has been taken quite seriously and has been investigated by the National Aeronautics and Space Administration and nuclear energy specialists for some years. One problem is the immense cost of ejecting materials into space. Costs are reduced somewhat if disposal is limited to long-lived wastes, but the state of the art in waste separation is such that sufficient removal of all long-lived hazards is doubtful. Only a marginal reduction might be possible.

If the cost of space disposal becomes less prohibitive, other problems are involved. Where should the waste be aimed? Since space is not empty, it has been suggested that the radioactive wastes be allowed to fall into the sun where they would be transformed in the extreme temperatures there.

Wastes could also be placed in orbit between the earth and Venus. Shuttles would place a waste package and orbiting transfer vehicle in an earth orbit. The waste could be coupled to the transfer vehicle and maneuvered into an orbit around the sun at a safe distance from the earth. The possibility of using a moon crater for retrievable storage of nuclear wastes is also being examined.

But what if an accident occurred on the launching pad, or before the wastes were a great distance from the earth? The risk of uncontrolled spreading of large amounts of highly radioactive material in the atmosphere makes this method of disposal unpopular. With costs in the range of thousands of dollars per kilogram of payload, one need not worry about space accidents with radioactive waste payloads in the near future.

If radioactive waste can be controlled so that small amounts are added to the natural background radiation, and if the waste is distributed all over the earth, is this a good way to deal with the problem? Dispersion has some disadvantages, even if it became technically practical. Some kinds of soil tend to trap and accumulate certain radioactive elements and will therefore concentrate them even if the original spreading is even. Plutonium is one such element. Erosion might place some of the concentrated wastes in the earth's water system so they would eventually reach the water supply for plants and animals, including humans.

What would happen in the ocean's food chain if radioactive wastes were spread around the world in minute but uniform concentrations? Tiny fish eat large quantities of algae, larger fish eat the tiny fish, still larger fish eat them, and so on. At each level, the concentration of radioactive materials increases many times, with a large accumulation by the time the fish are eaten by humans. This would make dispersal impossible even if technology could solve the problem of scattering waste in a uniform manner.

Disposing of wastes in Antarctica is a popular suggestion only because it is not in anyone's backyard. It has been suggested that high-level waste canisters be placed in Antarctic ice where heat from the wastes would melt the ice. If the ice is 3000 meters thick, the time for meltdown of the canisters to bedrock would be 5 to 10 years.

If canisters are lowered with cables attached, it will be possible to retrieve the wastes from where they are stored. Present estimates consider them safely retrievable after 200 to 400 years. Using a surface approach with pilings and cooling drafts of air, it might be possible to maintain a facility above the ice for about 400 years before the waste melted down into the ice.

Distance from population makes disposal of radioactive wastes in or on Antarctic ice attractive, but there are many problems. One is that an international treaty forbids it. And since the movement of ice over a period of many years is not clearly understood, ice disposal has been ruled out.

Recycling wastes has already been discussed.

The recycling of spent fuel rods provides more energy from the original fuel but it does not solve the waste problem, since the reprocessing procedure that makes recycling possible produces highly radioactive wastes itself. Also, many wastes cannot be recycled.

The ocean has long been considered as a repository for nuclear wastes. The ocean is no one's backyard—or everyone's. The load of wastes carried to the seas is already vast: attempts to dispose of nuclear wastes in the ocean depend on isolation that prevents circulation of wastes. A 1972 international agreement prohibits the dumping of high-level waste at sea, but a permit can be secured if it can be shown that wastes will not migrate.

Where does a safe disposal site exist? Parts of the ocean floor have not changed for 50 million years; a number of oceanographers believe that burial in these parts of the ocean floor may be superior to burial in land sites.* Three years of research indicate that some kinds of clay in these regions have the potential to isolate wastes from the ocean and from man.

G. Ross Heath, professor of oceanography at the University of Rhode Island, points out that the oceans include the least valuable real estate

*"High-level Wastes in the Seabed?" *Oceanus* (20, no. 1), Winter 1977 (entire issue).

on earth and that some of the areas beneath the seas are among the most stable parts of the earth's surface. They lie far from the regions where new sea floor is being created at the mid-ocean ridges and is being destroyed at the deep-sea trenches, and are not earthquake-prone. Nor do large rocks move in the most stable areas where sediments have accumulated for tens of millions of years. Such areas have obvious geological attractions. But will the radioactive wastes escape to the sea floor? What is the effect on the sediments of the heat generated by the waste? How can the wastes be placed on the sea floor? Suppose they are just dropped into the sea without trying to isolate them from the surrounding waters. This has been and is being done by some European countries with the hope that the low-level wastes will be so diluted as not to be harmful. It was reported at a conference on problems of the sea in 1970 that every living thing on and under the sea was being poisoned by radioactive wastes. Deformed backbones in embryo fish, for example, were found in the Irish Sea, which is polluted by the Windscale nuclear power station on the British coast. Radiation effects have been found in many marine organisms.

It has been calculated that the waters of the oceans are not vast enough to spread all of the wastes from military and industrial sources without contamination beyond safe limits for more than 20 or 30 years. No one can accurately predict how the wastes would be spread, how they would enter the food chain, or how long they would remain on the seabed. So disposal by dispersal in the seas is unsafe.

Suppose wastes are placed in a canister and dropped from a ship into the ocean. Will the canister survive hundreds of thousands of years in a corrosive marine environment, heated by the wastes inside? Since permanent storage with no leaks is necessary to prevent fouling the oceans, this method of disposal is quite controversial.

One of the main tasks of the seabed disposal program of the Energy Research and Development Administration is to find out how fast the radioactive wastes buried in the seabed would move up through the sediment and enter the environment of living things.

An important barrier being studied for land and seabed burial is the form of the waste material. Fusion into glass or ceramics is popular. When liquid waste is solidified by evaporation of the acid in it and is fused into glass, radioactive waste is easier to handle and insoluble in water. Plans call for the incorporation of wastes into a glass material much like that used in ordinary glass bottles in a ratio of about 75%

glass to 25% wastes. It is estimated that glass-incorporated wastes will be released to the surrounding environment a million times as slowly as liquid or soluble wastes, but the glass barrier is expected to decompose from the heat within about a thousand years.

Wastes incorporated into glass form would be sealed in noncorrosive titanium and zinc alloy canisters, each 1 by 10 feet. These canisters, a second barrier to migration, are expected to last a few thousand years.

Canisters would be buried in claylike ooze that covers the ocean floor in regions that are geologically quiet. They would be dropped from winch-equipped ships and would force their way 30 meters below the floor before coming to rest. Another possibility is pre-drilled holes in the sediment that hold two or three canisters. Less likely is the use of unmanned undersea crawlers to dig burial trenches in which the canisters are placed. No matter what method is used, the most important barrier to migration of radioactive wastes is the sediment.

Studies by oceanographers in the seabed emplacement program of ERDA indicate that some radioactive elements interact with clay grains and this slows migration. Research indicates that thorium would take 10 billion years to travel 1 meter in north Pacific clay. Work is being done on other radioactive waste elements to determine how fast they migrate.

Basing their estimates on laboratory tests, scientists believe that radioactive chlorine can be kept within a 100-meter radius for a million years, and radioactive cesium and strontium within a 30-meter radius for a million years; 3 inches of sediment will protect nuclear wastes from disturbances by most ocean organisms.

Where can the conditions needed for nuclear graveyards be found in the ocean? One area being studied is 600 miles north of Hawaii. The area has not yet been chosen as a disposal site, but a core taken from the area shows that sediments have remained undisturbed there for over 65 million years.

Seabed disposal of radioactive wastes has not yet been established as safe, but oceanographers have found no evidence of insurmountable technical difficulties. Gaps in knowledge still exceed the facts.

Geologists who have been exploring sites for land repositories are less optimistic. As in the oceans, a series of barriers is important. Again, the first barrier is embedding in glass. The radioactive waste inside the glass may have an important effect on structure, and water does dissolve slowly in glass. Estimates of time for leaching from glass range from 10,000 years after burial to indefinitely long. The glass would be cast in stainless steel cylinders and stored near or above the ground for at least fifty years after extraction from the reactor. The heat from the wastes is so high before then that above-ground storage or air-cooled repositories are necessary. The cylinder acts as a second barrier.

Once the radioactive wastes are buried, the geological barrier becomes important. Experiments have been carried out to determine how fast radioactive wastes move through rocks, but there is controversy about the results of the experiments. One problem is the continuous activity of the earth itself. Permanent storage time is measured in hundreds of thousands of years. Scientists know that many important changes have taken place in the earth in a few thousand years. For example, the Sahara Desert was partly fertile 5000 years ago. The English Channel was dry 7000 years ago. There were active volcanoes in the central region of France less than 10,000 years ago. The Rhine ditch which began faulting and collapsing 30 million years ago is continuing to do so. Climatologists are predicting another ice age well within the time limit needed for safe storage of high-level radioactive wastes.

Geologists are searching for inactive areas of the earth. The driest possible rock will be chosen.

But no rock is completely impervious to water, and water is present almost everywhere.

Several kinds of rocks have been explored as possible sites for repositories. Clay is not very sensitive to faulting, but it contains water that may migrate. Hard rock, such as granite and gneiss, is very sensitive to fracture; limestone is also ruled out for this reason. A rock that does not have water running though it now may have water in the future, after changes occur in the earth's crust.

After 40 years of burial, the average temperature of the rock just above and below a canister would be increased about 140° C (284° F). What effect will this have on rock formations? No one knows. Salt transfers heat away from the canisters more readily, and this is one reason that burial in salt beds is a favorite approach to the problem of isolating radioactive trash. Salt mines have other advantages. When salt is subjected to pressure, it flows like plastic and makes a seal. Salt beds are widespread in the United States and in many parts of the world and the technology of salt mining is highly developed. Since salt is water soluble, any salt deposits are obviously in comparatively dry areas. Here and there, brine (salty water) is present. This would, according to theory, migrate toward hot waste canisters and eventually corrode them. The brine would not carry the wastes away from their burial site, but the rate of migration toward the human environment would be hastened with canisters corroded.

At one site being explored near Carlsbad, New Mexico, it has been estimated that if all water flowing above the salt beds were diverted into them, the salt surrounding the waste would be dissolved and the waste exposed after 40,000 years. Such diversion could only happen if there were a 350-meter vertical fault. The chances of this are minute even over hundreds of thousands of years, but there is concern about it among environmentalists.

A greater threat is posed by salt beds that were dug some time in the past in a search for oil. Records were not always kept of where drill holes were made.

From 1963 to 1970 research was carried out on disposal of radioactive wastes in salt mines and plans were announced for the construction in 1970 of a pilot plant near Lyons, Kansas. The proposed site was beset with problems almost from the beginning. There were technical questions about some bore holes made in the salt mines there and about the fact that another area 1800 feet away was going to be mined. Suppose there was a breakthrough into an area where radioactive trash was stored? Besides the questions of geologists, concerned citizens raised their voices against putting the wastes in their backyards. The Kansas site was abandoned and new plans were made for a pilot plant, this time at Carlsbad, New Mexico.

The Carlsbad Waste Isolation Plant is often referred to as WIPP. Plans called for disposal of low-level and intermediate-level wastes from the military program. Later the area was considered for high-level military wastes, too. And by March 1978 it was proposed for civilian high-level wastes on a limited scale. One plan for burial in New Mexico called for the spacing of 1000 spent fuel assemblies over an area of as much as 20 acres. A Department of Energy task force has indicated that the spacing of assemblies should be arranged to assure that heat generated in each acre is considerably below what is expected for a full-scale operating repository.

Target dates for full-scale repositories have been delayed until near the end of the century. One of a number of problems is resistance by citizens of New Mexico. Environmentalists have banded together here as in other areas where nuclear radiation is a source of concern. The Clearinghouse for Environmental Action and other groups have intensified their efforts to block the project near Carlsbad and to block plans for a burial ground for commercial low-level wastes at a site in northeastern New Mexico.

Senator Harrison Schmitt has observed that New Mexico, now called the Land of Enchantment, does not care to become known as the Nuclear Garbage Dump state.

A nuclear test site north of Las Vegas, Nevada, was being considered for a national storage site for spent fuel early in 1979. Radioactive elements would be sealed in concrete silos, and in deep wells and tunnels bored into the Nevada gravel.

In one recent year, commercial nuclear power plants in the United States contributed energy equivalent to that supplied by 425 million barrels of oil, or 120 million tons of coal. There is no question about the need for more energy, and there is no question about the necessity for permanent disposal of radioactive wastes. There are many questions about which solution to these problems is the best.

There may be 140 to 380 nuclear reactors producing radioactive wastes in the United States by the year 2000. Military programs and commercial reactors in other countries will be adding to the number of radioactive atoms in the world's trash. The search for a solution has been long. It should be intensified. For the safety of people on the planet today and for generations to come, a world alert and major action are needed to answer the question *What should be done with nuclear wastes?* and to make sure that radioactive trash is isolated. Greater individual awareness and increased action are needed to find a safe way to dispose of radioactive atoms at present and in the future. YOU CAN BE ON GUARD TO MAKE CERTAIN THAT YOU KNOW THE CONDITIONS UNDER WHICH RADIOACTIVE WASTES ARE STORED IN YOUR BACKYARD.

GLOSSARY OF ATOMIC LANGUAGE

ALPHA PARTICLE: A positively charged particle given off by certain radioactive materials. An alpha particle consists of two neutrons and two protons bound together; it is the same as the nucleus of a helium atom. Its penetration can be stopped by a sheet of paper, and is not dangerous to plants or animals unless the alpha-emitting substance has entered the body.

ATOM: A particle that cannot be divided by chemical means. Atoms are the building blocks of chemical elements, and the atoms of elements differ from each other. According to present theory, an atom consists of a relatively dense nucleus and a much less dense outer part where electrons move around the nucleus.

ATOMIC NUMBER: The number of protons in the nucleus of an atom; also its positive charge.

ATOMIC WEIGHT: The number of protons plus the number

of neutrons in the nucleus. A relative number: the atomic weight of hydrogen-1 is 1; of uranium-238, 238. Uranium-235, the fissile form of uranium, has three fewer neutrons in its nucleus than uranium-238.

BACKGROUND RADIATION: The natural radiation in the environment, including cosmic rays and radiation from naturally radioactive elements.

BETA PARTICLE: An elementary particle emitted from the nucleus of an atom during radioactive decay. A negatively charged beta particle is identical to an electron: a positively charged beta particle is called a positron. Beta particles are stopped by a thin sheet of metal, but they can cause skin burns, and beta emitters are harmful if they enter the human body.

BOILING-WATER REACTOR: A reactor in which water, used as both coolant and moderator, is allowed to boil in the core.

BREED: To form fissile atoms, usually as a result of neutron capture: breeding may be followed by radioactive decay.

BREEDER REACTOR: A reactor that produces and consumes fissionable fuel. The new fissile material is created by capture of neutrons in fertile materials.

CHAIN REACTION: A reaction that stimulates its own repetition. In a fission chain reaction a fissionable nucleus absorbs a neutron and fissions, releasing additional neutrons. These, in turn, can be absorbed by other fissionable nuclei, releasing still more neutrons. A fission chain reaction is self-sustaining when the number of neutrons released in a given time equals or exceeds the number of neutrons lost by absorption in nonfissioning material or by escape from the system.

CLADDING: Material used to cover nuclear fuel in order to

protect it and to contain the fission products formed during irradiation.

CONTROL ROD: A rod, plate, or tube containing a material that readily absorbs neutrons, used to control the power of a nuclear reactor. A control rod prevents the neutrons from causing further fission, by absorbing neutrons.

COOLANT: A substance circulated through a nuclear reactor to remove or transfer heat. Water, carbon dioxide, air, and liquid sodium are common coolants.

CORE: The central portion of a nuclear reactor; it contains the fuel elements and possibly the moderator.

CURIE: A unit of radioactivity approximately equal to the amount available per second from 1 gram of radium. A curie is equal to 37 billion disintegrations per second.

DECAY, RADIOACTIVE: Spontaneous transformation of one atomic form of an element into a different form, or into a different energy state of the element. This results over a period of time in a decrease in the number of radioactive atoms in a sample. It involves the emission from the nucleus of alpha particles, beta particles, or gamma rays; or fission.

ELECTRON: An elementary particle with a negative electrical charge. Its mass is 1/1836 that of a proton.

ENRICHED MATERIAL: Material in which the percentage of a variety of atoms present has been artificially increased above the natural percentage level. Enriched uranium contains more fissionable uranium-235 (and less uranium-238) than natural uranium.

FAST BREEDER REACTOR: A reactor that operates with fast neutrons; it produces more fissile material than it consumes.

FERTILE MATERIAL: A material, not itself fissionable by slow-moving neutrons, which can be converted into fissile material by irradiation in a reactor. Two basic fertile materials are uranium-238 and thorium-232. When they capture neutrons, they are partly converted into plutonium-239 and thorium-233, which are both fissile.

FISSILE: Capable of fission.

FISSION: Splitting of atomic nuclei into two more or less equal parts with the release of energy and generally of one or more neutrons. Fission can occur spontaneously, but is usually caused by absorption into the nucleus of gamma rays, neutrons, or other particles.

FUEL CYCLE: A series of steps including mining, refining, fabrication of fuel elements, their use in a reactor, chemical reprocessing to recover the fissionable material remaining in spent fuel rods, reenrichment of the fuel material, and refabrication into new elements. A "throwaway" cycle ends with the storage of spent fuel rods.

FUEL REPROCESSING: The processing of reactor fuel to recover unused fissile material.

GAMMA RAYS: High-energy, short-wavelength electromagnetic radiations. Gamma rays frequently accompany alpha and beta emissions, and always accompany fission. They are very penetrating, and can be stopped best by dense materials such as lead or depleted uranium. Gamma rays are similar to X rays but usually have more energy and are nuclear in origin.

HALF-LIFE: The period of time in which half of any amount of radioactive substance will lose its radioactivity.

HEAT EXCHANGER: A device that transfers heat from one liquid or gas to another or to the environment.

HEAVY WATER: Water containing significantly more than

the natural proportion of heavy hydrogen (deuterium) atoms to ordinary hydrogen atoms. Heavy water is used as a moderator in some reactors because it slows down neutrons effectively and has a low probability of absorbing them.

HEAVY WATER REACTOR: One that uses heavy water as its moderator. Heavy water is such a good moderator that unenriched uranium can be used as a fuel.

IONIZING RADIATION: Any radiation that displaces electrons from atoms or molecules and thus produces ions. Examples of ionizing radiation are alpha, beta, and gamma radiation, and X rays. Ionizing radiation can produce severe skin or tissue damage.

LIGHT WATER: Ordinary water as distinguished from heavy water.

LIGHT WATER REACTOR: One that uses ordinary water as a moderator or coolant. Pressurized-water and boiling-water reactors use light water.

MODERATOR: A material, such as ordinary water, heavy water, or graphite, used in a reactor to slow the speed of neutrons and increase the probability of fission.

NEUTRON: An unchanged particle with a mass slightly more than that of a proton. Neutrons are found in the nucleus of every atom but hydrogen. A free neutron is unstable and has a half-life of about 13 minutes. Neutrons are used to sustain fission in nuclear reactors.

NUCLEAR REACTOR: An atomic device, in which a chain reaction can be begun, maintained, and controlled, for splitting atoms at a controlled rate. It consists essentially of a core with fissionable fuel, and usually has a moderator, a reflector, shielding, coolant, and control mechanisms.

NUCLEUS: The central part or core of an atom. It contains

protons and neutrons. It must be split to release atomic energy.

NUCLIDE: A general term used for all atomic forms of an element.

PARTICLE: A minute unit of matter which generally has a measurable mass. The main particles considered in radioactivity are alpha particles, beta particles, neutrons, and protons.

PLUTONIUM: A heavy, radioactive, man-made metallic element. Fissionable plutonium-239 is produced by neutron irradiation of uranium-238 and is used for reactor fuel and in weapons.

PRESSURE VESSEL: A strong-walled container housing the core of most types of reactors built to produce power.

PRESSURIZED-WATER REACTOR: A reactor in which heat is transferred from the core to a heat exchanger by water that is kept under high pressure to achieve high temperatures without boiling. This is called the primary system.

PROTON: An elementary particle with a positive electrical charge. All nuclei contain one or more protons.

RAD: *R*adiation *A*bsorbed *D*ose: The amount of radiation that gives an energy absorption of 100 ergs of energy per gram of substance (tissue). Nearly the equivalent of the roentgen, but the roentgen was designed to be used with X rays and gamma rays.

RADIOACTIVE WASTE: Equipment and materials that are radioactive from nuclear operations and for which there is no further use.

RADIOACTIVITY: A property of atomic nuclei of certain elements which break apart bit by bit, releasing alpha, beta, and gamma rays in a constant pattern.

REACTOR COOLING POND: A body of water in which reactor fuel elements are placed immediately after being taken out of the reactor core to absorb radioactive fission products. Elements are transported from the cooling pond to storage or reprocessing plants.

REM: *R*oentgen *E*quivalent *M*an: A unit devised to express the degree of biological injury caused by radiation. The amount of radiation that produces the same biological injury in humans as that resulting from the absorption of 1 roentgen of X-radiation.

RESEARCH REACTOR: A reactor primarily designed to supply neutrons or other ionizing radiation for experimental purposes such as training, materials testing, and production of a variety of radioactive materials.

ROENTGEN: A unit of exposure to ionizing radiation. The amount of gamma or X rays required to produce ions carrying 1 electrostatic unit of electrical charge in 1 cubic centimeter of dry air under standard conditions. Named for Wilhelm Roentgen, the discoverer of X rays (1895).

SAFETY ROD: A control rod used as a standby to shut down a nuclear reactor rapidly in case of an emergency.

SPENT FUEL ROD: A nuclear reactor element that has been irradiated to such an extent that it can no longer effectively sustain a chain reaction.

THERMAL NEUTRON: A neutron that has been slowed down by a moderator to an average speed of about 2200 meters per second at room temperatures from the much higher speed it had immediately on being expelled by fission. Thermal neutrons are slow compared with their original speed but not slow compared with common movement.

THORIUM: A naturally radioactive element. Fertile thorium-232 is abundant and can be transmuted to fissionable uranium-233 by neutron irradiation.

TRANSURANIC ELEMENTS OT TRANSURANIUM ELEMENTS: Elements that have atomic numbers greater than that of uranium. They are produced artificially and are important in nuclear wastes because of their radioactivity.

URANIUM: A radioactive element found in natural ores with an average atomic weight of approximately 238. The two principal forms are uranium-238 (99.3% of natural uranium), which is fertile, and uranium-235 (0.7% of natural uranium), which is fissile.

SOURCES OF FURTHER INFORMATION

Information about disposal of radioactive wastes is difficult to evaluate. The following excerpts are taken from U.S. government reports that express different points of view. When reading books such as those listed in "Suggestions for Further Reading" (page 107), one must consider whether or not the author is presenting opinions or slanting statements from the point of view of scientists who are for or against nuclear power.

T. A. Nemzek, Director, Division of Research Reactor and Development, U.S. Energy Research and Development Administration:

> Of course, though the waste material is small in size, this material is potentially hazardous and must be isolated from the environment—and it will be. There are several viable options for safely storing the wastes for many decades with little required supervision and for permanently disposing of

the material in underground geological formations which have been stable for millions of years. ["Nuclear Power—Myths and Reality," Address, August 6, 1975.]

William A. Radlinski, Acting Director, U.S. Geological Survey:

As a result of . . . expanded examination, modified concepts of geological disposal have evolved, and aspects of some older concepts have been questioned . . . . Because the authors are confident that acceptable geological repositories can be contructed, this paper should not be construed as an attempt to discredit the concept of geological containment or the work done in the 1960's and early 1970's. However, the earth-science problems associated with disposal of radioactive wastes are not simple, nor are they completely understood. The many weaknesses in geological knowledge noted in this report warrant a conservative approach to the development of geological repositories in any medium. Increased participation in this problem by earth scientists of various disciplines appears necessary before final decisions are made to use repositories. Basic philosophical, as well as technological, issues remain to be resolved. [Foreword, "Geological Disposal of High-Level Radioactive Wastes—Earth-Science Perspectives" (1978), Geological Survey Circular 779, free from: Branch of Distribution, U.S. Geological Survey, 1200 S. Eads St., Arlington, VA 22202.]

Comptroller General, Report to Congress, September 9, 1977:

Growth of nuclear power in the United States is threatened by the problem of how to safely dispose of radioactive waste potentially dangerous to human life. Nuclear power critics, the public, business leaders, and Government officials concur that a solution to the disposal problem is critical to the continued growth of nuclear energy.

Radioactive wastes being highly toxic can damage or destroy living cells, causing cancer and possible death depending on the quantity and length of time individuals are exposed to them. Some radioactive wastes will remain hazardous for hundreds of thousands of years. Decisions on what to do with these wastes will affect the lives of generations to come.

To safeguard present and future generations, locations must be found to isolate these wastes and their harmful environmental effects. A program must be developed for present and future waste disposal operations that will not create unwarranted public risk. Otherwise, nuclear power cannot continue to be a practical source of energy.

The General Accounting Office found:

- public and political opposition to nuclear waste disposal locations;
- gaps in Federal laws and regulations governing the storage and disposal of nuclear waste;
- geological uncertainties and natural resources tradeoffs encountered when selecting "permanent" disposal locations;
- lack of Nuclear Regulatory Commission regulatory criteria for orderly waste management operations, such as solidification of waste, designing proper waste containers, and transporting nuclear waste;
- overly optimistic schedules for demonstrating the safety of the Energy Research and Development Administration's proposed waste disposal locations and waste management practices; and
- lack of demonstrated technologies for the safe disposal of existing commercial and defense high-level waste.

Now that commercial reprocessing of spent fuel has been indefinitely deferred, finding solutions to problems in storing and/or disposing of nuclear spent fuel will become a top priority matter. ["Nuclear Energy's Dilemma: Disposing of Hazardous Radioactive Waste Safely," Energy Research and Development Administration, Nuclear Regulatory Commission.]

SUGGESTIONS FOR FURTHER READING

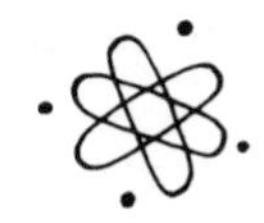

Beckmann, Peter, *The Health Hazards of Not Going Nuclear.* Boulder, Colo.: Golem Press, 1976.

Brockris, J. O'M., *Environmental Chemistry.* New York: Plenum Press, 1977.

Commoner, Barry, ed., *Radioactive Contamination.* New York: Harcourt Brace Jovanovich, 1975.

Francis, John, and Paul Albrecht, *Facing Up to Nuclear Power: Risks and Potentialities of Large Scale Use of Nuclear Energy.* Philadelphia: Westminster Press, 1976.

Inglis, David Rittenhouse, *Nuclear Energy: Its Physics and Its Social Challenge.* Reading, Mass.: Addison-Wesley, 1973.

Keeny, Spurgeon M., Jr., *Nuclear Power: Issues and Choices.* Report of the Nuclear Energy Policy Study Group. Cambridge, Mass.: Ballinger, 1977.

Lowrance, William, *Of Acceptable Risk: Science and the Determination of Safety.* Los Altos, Calif.: William Kaufmann, 1976.

Maxwell, Kenneth E., *Environment of Life*. Encino, Calif.: Dickenson, 1973.

Munson, Richard, ed., *Countdown to a Nuclear Moratorium*. Washington, D.C.: Environmental Action Foundation, 1976.

Panel on Land Burial, Committee on Radioactive Waste Management, *The Shallow Burial of Low-Level Radioactively Contaminated Solid Wastes*. Washington, D. C.: National Academy of Sciences–National Research Council, 1976.

Webb, Richard E., *The Accident Hazards of Nuclear Power Plants*. Amherst: University of Massachusetts Press, 1976

Wilson, Carroll, *Energy: Global Prospects 1985–2000*. New York: McGraw-Hill, 1977.

The following publications are available as a consumer service. One copy of each of several titles may be requested and will be sent free of charge from: U.S. Department of Energy, Technical Information Center, P.O. Box 62, Oak Ridge, TN 37830.

Fusion. 1977. 9 pages. Explains a process similar to that by which the sun generates energy, and that could be used for future electrical energy production.

The National Energy Plan. 1977. 104 pages. The plan submitted to Congress in April 1977 for energy management, production, research, pricing, and conservation into the mid-1980's.

Nuclear Power Plant Safety. 1976. 6 pages. Describes the three levels of safety incorporated into the design of reactors.

Plutonium in the Environment. 1976. 10 pages. Describes plutonium production, benefits of its use, and the various ways it might affect the environment.

Safeguarding of Nuclear Materials. 1976. 8 pages. Describes the basic components of the nuclear system, including facility protection and inventory control.

Shipping of Nuclear Wastes. 1977. 10 pages. Tells how radioactive wastes are packaged and transported, including the safety precautions employed.

Tips for Energy Savers. 1977. 44 pages. Offers practical and simple ways to save energy in the kitchen, workshop, garden, and car.

Fact sheets on alternative energy sources are available at no cost from the same address listed above. Each fact sheet was written in 1977 and is 4 to 8 pages long.

Alternative Energy Sources: A Bibliography
Alternative Energy Sources: Environmental Impacts
Alternative Energy Sources: A Glossary of Terms
Breeder Reactors
Conventional Reactors
Electricity from the Sun I (Solar Photovoltaic Energy)
Electricity from the Sun II (Solar Thermal Energy Conversion)
Energy Conservation: Homes and Buildings
Energy Conservation: Transportation
Energy Storage Technology
Fuels from Plants (Bioconversion)
Fuels from Wastes (Bioconversion)
Geothermal Energy
New Fuels from Coal
Nuclear Fusion
Solar Heating and Cooling
Wind Power

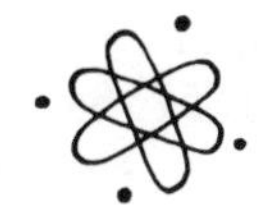

INDEX

ABOUT THE AUTHORS

Margaret O. Hyde is the author of an outstanding list of science books for young readers. Mrs. Hyde received her master's degree from Columbia University and received an honorary doctor of letters degree from Beaver College, her alma mater. Her recent books for McGraw-Hill include *Know About Alcohol; Addictions: Gambling, Smoking, Cocaine Use and Others;* and *Brainwashing and Other Forms of Mind Control.* She is the mother of Bruce G. Hyde. Margaret Hyde and her husband live in Burlington, Vermont.

Bruce G. Hyde received a bachelor of science degree from Lehigh University and a master's in resource economics from the University of Vermont. He is currently assistant director of community development for the city of Burlington, Vermont. He and his wife live in South Hero, Vermont.

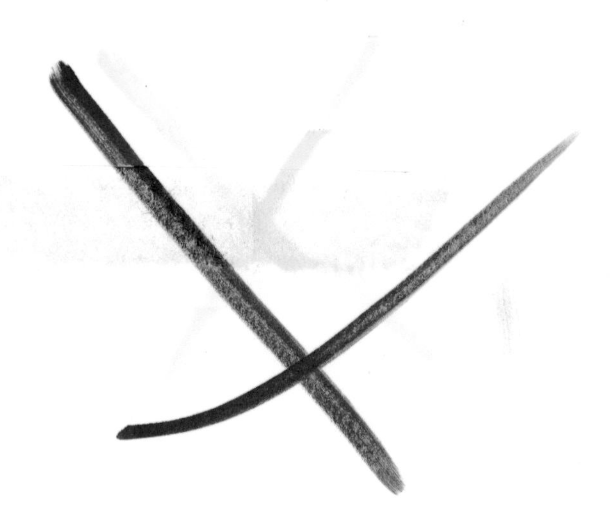